The **Four Components**

of a **Fast-Paced**

Organization

Going Beyond Lean Sigma Tools

Robert Baird

The Four Components of a Fast-Paced Organization

Going Beyond Lean Sigma Tools

CRC Press
Taylor & Francis Group
Boca Raton London New York

CRC Press is an imprint of the
Taylor & Francis Group, an **informa** business

A PRODUCTIVITY PRESS BOOK

CRC Press
Taylor & Francis Group
6000 Broken Sound Parkway NW, Suite 300
Boca Raton, FL 33487-2742

International Standard Book Number-13: 978-1-4822-0600-5 (Hardback)

Library of Congress Cataloging-in-Publication Data

Baird, Robert, 1953-
   The four components of a fast-paced organization : going beyond lean sigma tools / Robert Baird.
      pages cm
   Includes bibliographical references and index.
   ISBN 978-1-4822-0600-5 (hardback)
   1. Leadership. 2. Mentoring. 3. Organizational effectiveness. 4. Knowledge management. 5. Six sigma (Quality control standard) 6. Lean manufacturing. I. Title.

HD57.7.B344 2013
658.4--dc23
                                       2013036999

Visit the Taylor & Francis Web site at
http://www.taylorandfrancis.com

and the CRC Press Web site at
http://www.crcpress.com

Contents

Foreword

In my role as vice-president of North American operations for a global manufacturer, I would frequently be asked if we were a world-class manufacturing organization. I would answer yes, as we had a dynamic market share, state-of-the-art product offerings, and a seasoned management team. I really did not understand exactly what *world class* meant in detail, but assumed we were. From interfacing with other senior managers outside of my organization, I learned what world-class manufacturing really was and realized we were far from it.

We then embarked on a transformation to world-class, Lean, Six Sigma—many names are used, but the essence is developing a fast-paced organization that goes beyond Lean Sigma tools and emphasizes leadership, organizational structure, process design, and knowledge sharing.

Robert Baird was our quality manager at the time we started this transformation. He had recently joined us, and he embraced our evolution to a world-class organization, leading the way. Robert is a Six Sigma Black Belt and has detailed knowledge of both the hard, technical Six Sigma skills and the soft, people and organizational development skills.

Robert was instrumental in developing our self-directed, high-performance teams throughout the plant that enabled 100% of our workforce to be engaged and committed to executing our manufacturing strategies.

Robert led our Management Steering Team, where he enabled a horizontal flow of knowledge and collaboration between departments, eliminating the departmental silos that tend to develop in an organization, leading to fast-paced communications and problem solving within and between departments.

This book demonstrates the ability to simplify the organizational structure with leaders as supporters, teachers, and promoters improving the speed of transformation to a Lean environment.

My experiences as a senior manufacturing manager who participated in this transformation to Lean exceeded my most optimistic expectations. Our yields steadily increased to 99%+ in key product lines, our productivity improved by 50%, indirect labor costs were cut in half, and we unleashed the potential of our workforce. I encourage organizational leaders interested in transforming to world class to read Robert's book detailing how your manufacturing operations can become an elite, value-adding part of your overall business.

Louis Bisasky
Vice President of North American Operations

Acknowledgments

There are certain people in business and life you have the good fortune of knowing. My understanding is that there are very few people who have the inherent passion and ability to motivate, whatever the situation. They are passionate people who are able to see the best in people, where job position plays no role and company politics are a negative input.

To my wife: It is crazy that I have such a loving and smart person and am always asking more from her. She always helps and her love is endless.

To the employees of the Owings Mills, Maryland Advanced Card Center: The inspiration for this book came from your ability to achieve such a high level of performance. A once-in-a-lifetime experience and rewarding journey—thank you so very much!

To my friend and mentor Lou Bisasky: The Lean journey we traveled in Owings Mills was led by Lou. He is a true leader who valued the people who worked with him. To this day I have never met another leader with his many leadership qualities. Lou ensured it was fun.

To my friend Jim Jamieson, who is one of the rare people who realizes the importance of empathy and having 100% commitment: He is passionate, caring, and motivated to accomplish world-class business results. He always globally promoted Owings Mills as a world-class operation.

Finally, to Pat Callahan, my brother-in-law, who was recently diagnosed with the terrible disease of amyotrophic lateral sclerosis (ALS): He maintains a wonderful outlook on life.

About the Author

Robert Baird has practiced successful applications of Lean Sigma globally for the past 20 years, including three complete Lean Sigma transformations. Each transformation achieved world-class business results.

Baird worked for Schlumberger Oilfield Services for 20 years, holding various management positions. He then transferred to Gemalto and worked there for 14 years, holding top management positions including vice president of operations and global world-class enterprise manager. Baird is currently founder and president of Lean Teams USA consulting firm.

He has introduced and supported a global Lean Sigma strategy for business results with his work in countries including Brazil, Canada, China, Finland, France, Mexico, Poland, Singapore, the United Kingdom, and the United States.

Introduction

The greatest enemy of the best is the acceptance of good.

Paraphrased from Voltaire

If you look at the characteristics of a successful organization, you will find speed as one of them. Staying ahead and achieving business results at a pace faster than the competitors are a distinct advantage. Within Lean we have some focus on just-in-time, lead time, and on time delivery—it is the nature of what Tom Peters (1992) called "Nanosecond Nineties"—but what are we doing about it? Are organizations focused on improving their organizational speed? We found that when there is a focus on speed, industry-leading improvements come with it and people are motivated to keep going; momentum is created, and all employees are engaged to execute the strategy. These are the outputs or effects on an organization when implementing all four of the components of a fast pace. Implementing one, two, or three of the components will certainly achieve results but not the sustained world-class results we are looking for. We all want a culture of continuous improvement, learning, and customer orientation; and this is what the four components were designed for. Following the implementation steps will inherently develop all of these organizational characteristics. It starts with the Leadership and Mentoring component. The leaders must be on board first, and then each of the other three components are to follow. The Leadership and Mentoring component provides the base that sustains the momentum and starts the culture. Leaders take responsibility for developing people who are trained, motivated, and supported to identify, solve, and fix problems. Leaders must empower people to be capable of process ownership. Leaders must guide and support a production system of continuous flow and quality products and service. Finally, leaders must build a learning organization. They are responsible for operational excellence.

Achieving excellence comes with a number of challenges. The first is to recognize the need for change. There have been many organizations with mediocre results from a Lean Sigma transformation, and I propose it is mainly from not providing a complete organizational change. Implementing Lean Sigma tools as the base will not provide world-class results. It all starts with leadership understanding the scope of the required change. This does not mean it is a monumental and daunting change. With a leadership style of participation and learning, the change does not have to be daunting. It does involve the understanding and knowledge to develop and engage 100% of the workforce and ensuring 100% of the workforce clearly knows the strategy and how they can contribute to the strategy. Excellence also requires an efficiently designed process, a support organization, and systems to provide fast knowledge sharing.

In this book I describe, with implementation detail, four key components of an organization that must be in place to reach and sustain world-class business results at a fast pace. I also wrote the book in a style to provide only the needed details. How many people have read a 300+ page book and have only come away with one or two ideas they could use? My argument is for you to be able to get results from these proven components with, as you will see, reading as few pages as possible.

I was involved in a very special Lean Sigma transformation at an organization going through a market alteration. At the time there was discussion of outsourcing our production lines to low-cost countries; you could say it was our "burning platform." The president of the organization knew the objectives were challenging and decided to take a very different approach. The main objective (later to be communicated as our single focus) was productivity above the proposed low-cost countries. The first step he took was to hold a meeting with all of the key managers, where he was very candid in presenting the daunting objective set by corporate. He also said that we had to meet this objective within a year. The next step was to brainstorm where we could possibly improve productivity—nothing was off the table. The next step was to determine where the "low-hanging fruit" was, and what we could do to provide the pace in achieving the results. At this time Lean Sigma was suggested as a methodology to support meeting our objectives. At the time only the human resources manager and I had experience with Lean Sigma. We agreed to hire a consultant to get us started. The management staff was then trained in 5S, Total Productive Maintenance (TPM), Kaizen, and manufacturing cells.

The president was not confident that these tools alone would meet our objectives, so he got us back together again and explained that we needed something else. The suggestion of self-directed teams came up. The president thought about it and agreed. At this same meeting we also decided on a single-focus strategy of productivity. The president also decided that we needed to have this group of managers meet every second day to create the required pace for achieving the objectives on time. We then decided to call ourselves the Management Steering Team to provide some structure on how we would govern; we needed the input of everybody! From then on we always had an agenda item to discuss how we were creating pace toward our business results, and this is how the four components were developed. The result was that we doubled productivity, and along with these objectives, yield improved 75% and on-time delivery above 90%.

So here are the four key components:

1. Leadership and Mentoring
2. Process Design and Visual Value Streams
3. Organization Structure for Sustainment
4. Fast Knowledge Sharing

I also provide some explanation of some key Lean Sigma tools because they are critical in achieving the results you are looking for. However, Lean Sigma tool implementation by itself has been the mistake of many organizations. Too often the pattern is that someone in the organization decides Lean Sigma is something they would like to try. They start with some training, usually 5S, TPM, or even completing a value stream map. They see the early results and conclude: We were right—it did work. Then a few months later even the Lean tools are not part of the culture. This is a huge waste.

The approach I describe here with the four key components is more of a leadership and culture approach. I say leadership first because the correct culture cannot be established without talented leaders. There must be a strong leader, and this leader understands the power of teams, has empathy, is unpretentious, takes calculated risks, has fun, and is a strong communicator. The leader understands that Lean Sigma will be a complement to achieving business results; but he or she clearly knows that highly developed and motivated people, a self-directed team organization, and fast knowledge sharing and learning will bring sustained business results. It is also understood that this culture must be realized at a fast pace to maintain momentum. Leaders must understand that they must go to the Gemba; it is their responsibility to support, teach, and promote (what I call STP) the single focus strategy to achieve world-class results.

The single-focus strategy needs some explanation, as some people might think it is a limitation toward achieving world-class results. It is, as it states, a single focus; and the selection process for this measure is absolutely essential. We all receive what we need to improve from somewhere, from our customers first, from our leaders, or maybe from our employees. And yes, it is never one measure we are summoned to improve. However, I have seen managers who had more than 100 measures they were trying to improve and actually reviewed each one on a quarterly basis. Do you think they were successful? The single-focus strategy makes you think about the one measure that, if improved, would most satisfy your customers, both external and internal. A great example is lead time. You cannot have world-class lead time with poor quality or productivity or even high costs. These measures are actually inputs to the output of lead time. So you must improve the inputs to achieve world-class lead time. However, it is now much easier to communicate to everyone what the organization's focus is and where you are going. It is also much easier for everyone to understand how they contribute toward this single focus. Understanding how you contribute to the success of the organization increases your individual value and thus motivation. The single-focus metric is visually managed so that people see progress on a daily basis and can have daily discussions on how they can further improve it.

The value streams must be of a design that is conducive to smooth flow. Work in progress (WIP) is low and the cells are in continuous flow. These value streams also require value-added visual management. The value comes from the managers being able to quickly see if the improvement projects related to the single focus are progressing, if orders are on time, if the quality is normal, if standard work is improving, and if gained results are being sustained. This design and visual management contribute significantly to the fast pace as the problems are easily seen, realized, and addressed. Small actions are assigned and followed up on a daily basis. Small improvements are made every day, in fact, every shift.

Organizational structure, also key to fast pace, provides a strong support structure for everyone. The vertical structures or silos are broken down when people are contributing more toward the single focus than to their own department measures. Department managers are offering resources to other departments when they need it. The structure starts with the Management Steering Team, and most managers are on this team with the main objective of strategy deployment. The structure of the organization includes the team facilitator, who used to be the supervisor; and the self-directed teams, that are now empowered to do most of the tasks that were previously performed by the supervisor.

Finally we get to the component of Fast Knowledge Sharing, which uses methods and tools like worldwide accessible databases, process standardization, social networking, and breaking down of department silos. Being fast paced in knowledge sharing minimizes resources required for process improvement, and because of this, these same resources can do more projects that contribute to fast-paced results.

The four components also ensure that results are sustained because of visual management, Gemba walks, the organization's structure, and process ownership.

One of the keys is to keep strong momentum by leaders being involved at the value stream on a daily basis. When you have the four components in place, the organization is fast paced, which provides the momentum.

REFERENCE

Peters, Tom. 1992. *Liberation Management: Necessary Disorganization for the Nanosecond Nineties*. New York: Knopf.

1

Component 1: Leadership and Mentoring

A fast-paced organization starts with leadership. In most organizations are leaders who understand the benefits of a fast pace, and if you are doing a bottom-up Lean transformation I would recommend that the leaders of your organization start by learning about this component, Leadership and Mentoring. Leaders have a responsibility to develop a strategy that will be successful; they are responsible for the organization's culture; they need to understand their role in developing people so that everyone can contribute to the success; and they must also understand they have a social responsibility. It is key that leaders are aware of various and proven success factors. It is part of their responsibility to provide mentoring, proven and productive value-stream designs, organizational structure providing support and transparency, and finally systems and programs to facilitate fast knowledge sharing. These are the four components for speed and operational excellence.

LEADERSHIP GOING FORWARD

A manager's primary tasks are planning, organizing, controlling, and leading. Whereas past leadership efforts have been all about the individual as a leader, presently it is about the collective efforts of a team. Today's leaders are coming across new challenges where there isn't any history to base an informed decision on, and leaders must look to their social networks, employees, and other leaders to help them make the best decision.

It's OK to show that you don't always have the answer. A good leader at the very least knows where to find that answer. Being vulnerable will provide competence, as people are more willing to show trust in a leader that shows some vulnerability.

An example of a long history of coming to a consensus is the U.S. Cabinet. The U.S. Cabinet comprises the most senior appointed offices. Established in Article II, Section 2, of the U.S. Constitution, the role of the Cabinet is to advise the president on any subject he may require relating to the duties of each member's respective office. Dating back to the first U.S. president, George Washington, the Cabinet has served the country well. Leadership is required, but before policy deployment, diverse knowledge needs to be a key input. Organizational leaders today require a strong team for guidance; decisions are by consensus.

What is needed today is *not* heroic figures who make great, inspiring speeches. The essence of Lean leadership is not who the leader is and what image he or she projects—the issue is about the leadership practice that accomplishes the aim, the purpose of the business. Any system that relies too heavily on charismatic "leadership" is inherently fragile and all-too-dependent on the individual who happens to be in charge. The real issue is how leadership builds systems that are the operational result of disciplined Lean practice—a problem-solving culture that creates continuous improvement that delivers business results while always solving customer problems. I must also mention here that leaders are also responsible for other duties like managing the day-to-day production needs, updating information systems, data analysis, managing the budget, and yes, pleasing the boss. This is not leadership but is the management side of their job. There is a difference between the leadership and management requirements of the job. Leadership is more about developing the required culture, developing a successful strategy, developing and empowering people, and adding individual value and security to the people they lead.

In the final analysis, how a decision is made at the strategic level is just as important as the decision itself. Also, the best decision in the world is nothing without a powerful consensus for action. The most perfect consensus in the world is useless unless it has produced a decision that is good for the organization.

At the front end of the entire team-decision-making process are the *inputs*. These are the key inputs for the business, and they must be more developed and of higher quality than the competitor's. All of the decision makers must be crystal clear on what these inputs are.

People who enter into consensus decision making must come armed with critical and creative thinking skills that will allow them to function efficiently and effectively at the strategic level.

The management team must use a process to cultivate their single-focus strategy. Winston Churchill once said, "The only thing worse than having allies is not having them." However, the glue that held the Anglo-American-Soviet alliance together during World War II was the determination to defeat Nazi Germany, fascist Italy, and Japan's military government reaching for control of East Asia. The single-focus strategy was the determination to defeat.

In 1980 the U.S. hockey team winning the Olympic gold was labeled "Miracle on Ice." This became a reality because of their coach, Herb Brooks, declaring the single-focus strategy of winning the Olympic gold when he was at the beginning stages of picking players. Many people in the sports world thought he was crazy and said he should realistically develop his strategy at maybe reaching the medal round.

Before a joint session of Congress on May 25, 1961, John F. Kennedy made the single-focus decision of a mission to the moon. This single-focus strategy united a nation. On July 20, 1969, Apollo 11 commander Neil Armstrong stepped off the ladder of the Lunar Module and onto the moon.

Robert Ballard started his single-focus strategy of finding the *Titanic* in 1973. Early in the morning of September 1, 1985, the *Titanic* was found.

All of these historic and memorable events came about because of a single-focus strategy. These people did not proclaim a complex vision. Their passion enabled them to communicate an outcome for everyone who followed to clearly understand the direction and hence how they made decisions. From this single focus many events had to be successfully executed, but the results were all clearly tied to meeting this single focus. Herb Kelleher, who was one of the founders of Southwest Airlines, maintained and based all of the organization's decisions on the single-focus strategy of "Southwest is *the* low-fare airline." He would tell his employees, "I can teach you to be CEO of Southwest Airlines in 5 minutes"—they just had to base their decisions on this single-focus strategy. He once had a flight attendant propose to offer a chicken-on-salad meal for one of their longer flights, and Herb simply asked if that would help meet the single-focus strategy, "Southwest is *the* low-fare airline." The meal was never implemented. Probably the most repeated single-focus strategy is Nike's "Just Do It." Within Nike this eliminated much wasted time in contemplation of meeting strategy objectives. Strategy deployment exercises became much shorter and things got done. I once worked for a large organization that had the single-focus strategy "zero injuries." This was an organization of 113,000 employees, so to help communicate the strategy the CEO declared that all meeting agendas, no matter the subject, had to start with a safety topic. He was the model of this policy, and it is now the culture of the organization.

Too many organizations come out, every year, with at least a half a dozen targets. They might have a vision statement proudly developed by a few leaders of the organization but tattered and forgotten. This confuses most of the workforce, so a high percentage of the organization do not base their decisions on meeting the strategy. Because they do not know the strategy, they do not know how to contribute. Because they do not know how to contribute, they have little individual value in being part of the organization's success.

With a single-focus strategy there is no complexity. What to do is easily cascaded throughout the organization. A high percentage of the workforce is now clear on what they have to do, and individual value is extremely high because they can see how their efforts contribute to the success of the organization. The complete organization has a clear line of sight to the single-focus strategy.

Leaders need to ensure they have a single-focus strategy to realize the significant benefits of an engaged workforce.

This method of determining the single focus becomes easier when the management team asks, "What, more than anything, do our customers want, that if we achieved it we would have a definite competitive advantage?" The answer common to most organizations is lead time improvement. To achieve this, the management team should use the Six Sigma tool of SIPOC (suppliers, inputs, process, outputs, customers). In this case the single output is lead time, and some of the inputs are quality, work in progress (WIP) levels, process capability, supplier deliveries, safety, and others. So you can now see how a single-focus strategy still requires a strategy of improving in other areas that are inputs to the single focus. This approach makes it much easier to communicate the strategy, why we are working on these inputs, and how everyone contributes.

A strategic leader can utilize decision-making teams as a powerful asset in successfully coping with the environment. Such teams improve their decision making by using a process of consensus, which is useful when developing a single-focus strategy. Knowing how to forge consensus for single-focus development and implementation is critical to successful management and leadership. The single-focus strategy does not prevent improvement in other areas of the business but enhances and provides speed to these other improvements because the culture of improvement is developed at a fast pace. Consensus decision making offers the benefit of using more fully the experience, judgment, perceptions, and thinking of a team of people.

A high-performing team can be a positive force in assessing strategic situations and formulating the single focus.

Effective strategic leaders employ a strategic team to help them in the formulating process. This team "sees" the strategic environment from various frames of reference, visualizing the effectiveness of proposed strategies over time. Teams help leaders to understand a complex situation and gain insight into how to achieve long-term objectives, allocate resources, and integrate operational and tactical decisions into strategic plans.

Strategic teams that perform with unity of purpose contribute to the creation of strategic vision, develop long-range plans, implement strategy, access resources, and manage the implementation of the strategy. Given the nature of the strategic environment and the complexity of both local and global issues, strategic leaders must use teams. They cannot do it alone.

The team's objective is to find the highest quality solution to a complex strategic problem and minimize risk of failure. In execution of the agreed strategy the management team must maintain the single focus. This approach will clarify any complications that arise during the planning and execution of the strategy.

With the single-focus strategy agreed upon, it is now the team's responsibility to determine the key metric and how this will be cascaded, communicated, and supported. The self-directed team's organization plays a critical role in meeting these targets through homogeneous improvement efforts throughout the value streams.

Communication of the single-focus strategy does not happen through a one-time organization meeting, posting, or newsletter. These are helpful, but the key communication is through the daily management Gemba walk, where questions related to the single-focus strategy are tested for comprehension. Does everyone know the single-focus strategy and how they are able to contribute? These are two critical questions to continue to ask until the management team feels comfortable that 100% of the people clearly know.

A study by Ernst & Young (1998) of 275 professional managers, "Measures That Matter" determined that the ability to execute strategy was more important than the quality of the strategy itself. Another study by management consultants reported that less than 10% of effectively formulated strategies were implemented successfully (Kiechel 1982). This is because of the typical organization's low level of involvement in directly executing the strategies, which results in lesser results even though the strategy is very well devised. With a self-directed team organization, a high percentage of people know the single-focus strategy and know how to contribute. This generates individual value and from this motivation to execute.

LEADERSHIP RESPONSIBILITY

The best minute I spend is the one I invest in people.

—**Kenneth Blanchard**

According to a study of more than 1,700 CEOs from around the globe, human capital was cited as the most important factor in maintaining competitive advantage (see Figure 1.1). So why do organizations continue

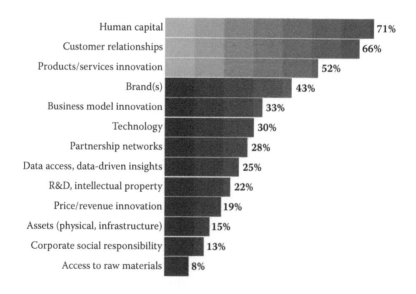

FIGURE 1.1
Human capital, most critical asset. (From *Leading Through Connections: Insights from the Chief Executive Officer Study*, IBM, 2012. Reprinted with permission.)

to remove this number one asset during a time when they feel they need cost reductions: Is it the only cost they have? Can they not provide innovation to generate more profitable revenue? Why do most companies spend 40% to 60% of their money on salaries and only 1%–2% on training? Is this all an organization gets from highly paid leaders?

In 1914 Henry Ford shocked the world by announcing he was going to double the salaries of most nondirect employees to $5 per day. After his announcement, thousands of prospective workers showed up at the Ford Motor Company employment office. They came from all over, including Europe. Ford not only doubled their salaries but also reduced their hours from nine to eight. This enabled three shifts, increasing productivity. Now the motivation to do this was that Ford Motor Company was experiencing problems with a high attrition rate. However, Henry also reasoned that with mass manufacturing and with fast cycle times resulting in lower cost, employees with higher wages could afford to purchase the products of their labor. Henry Ford changed manufacturing forever. He demonstrated responsibility to the world leadership and during the time was actually thought of as a folk hero. Where are the Henry Fords of today?

Leaders today must stop thinking of their employees as an expense and start thinking of them as an investment. Let's take a look at a sample income statement:

Revenue:	20,000
Cost of Goods Sold (COGS):	13,000 (direct labor is 4,000)
Gross Profit:	7,000
Selling Admin:	5,000
Depreciation:	300
Total:	5,300
Earnings before Interest and Taxes (EBIT):	1,700

Now let's say the organization is focused on improving EBIT. In most cases, what is the decision of most leaders today? You guessed it: Cut direct labor! But let's now take a look at improving any line on the income statement by 5% and understand the effect on EBIT. So we reduce labor by 5%, EBIT improves to $1,900, an 11.8% improvement. Now let's look at a 5% improvement in revenue and EBIT improves to $2,700, a 58.8% improvement in EBIT. Which one is simplest to do? Of course, cut the expense line of our most valuable asset, the people. Absolutely, to improve the revenue by 5% takes more innovation, creativity, better people than the competitors have, and leaders who can lead. Which company do you want to work for? If you are a leader, which direction and culture do you want to aim for?

On this topic of cost, the other method popular with leaders is outsourcing to lower-cost countries. They started by going to Mexico, China, and South America. Today China is not low-cost enough, and products are now going to countries like India and Cambodia. Over the past 15 years we have begun to see a change: U.S. manufacturing productivity has increased 4% per year. This is because some leaders are starting to look at the ignored costs when sending products to low-cost countries:

- Intellectual property at risk
- Cost and time of travel to visit distant plants
- Costs due to poor quality
- Negative impact of separating corporate engineering and other skills
- High oil prices and thus cost of shipping finished goods
- Customers demanding faster lead times, which can result in loss of revenue if not met
- Cost of natural gas—twice as high in China as in the United States
- Wages in China have been increasing 15% per year
- China's overheating economy, so service has increased in price
- Natural disasters interrupt key supply chains and then the ability to recover

"The re-shoring moves come as average manufacturing costs continue to fall in the U.S.," Jennifer Booton wrote in an article for Fox Business (2012). "BCG [Boston Consulting Group] estimates that they will be 8% lower than in the U.K. in 2015, 15% lower than in both Germany and France, 21% lower than in Japan and 22% lower than in Italy. China will still be about 7% cheaper than the U.S., but that doesn't include the high cost to ship bulk items around the world."

Many challenges for leaders have been dragged from previous years to the present. The main challenge is related to coaching people to be experts in whatever they do. Leadership has this responsibility to the people who work for them because these people are adding value to the products and services every minute of a working process. These people are clearly the executers of the value that the R&D and marketing departments have developed. One challenge from previous years is competing with lower cost countries. How are we going do to this? We have to do something lower cost countries cannot, and we must develop our people to a higher level than the competition is. Leaders today must stop thinking of their employees as an expense and start thinking of them as an investment.

At-will employment is a doctrine of American law that defines an employment relationship in which either party can immediately terminate the relationship at any time, with or without any advance warning and with no subsequent liability, provided there was no express contract for a definite term governing the employment relationship and that the employer does not belong to a collective bargaining group (i.e., has not recognized a union). Most leaders within U.S. organizations unfortunately take advantage of at-will employment. It is very easy to remove the most valuable asset and remain receiving a year-end bonus. Productivity, when defined as units produced ÷ direct labor, can be improved *simply* by removing direct labor. Lean Sigma, when starting to gain success, provides a myopic leader with the first thought and opportunity to remove direct labor. This is a fatal error. The same leaders created the beginnings of momentum to provide some success but not world-class results. The valuable people who took the risk, working within the process, then begin to lose confidence in the new Lean Sigma effort. Even after the basic building programs like 5S (see Chapter 5), Kaizen events, Total Productive Maintenance (TPM), and some continuous flow from cell layouts, they lose most of the required momentum. And then fire fighting comes back also.

Leaders fail when they cannot improve productivity without taking advantage of things like the at-will employment doctrine of American law. The challenge for leaders is not to remove our most valuable asset but to use known Lean Sigma methods of improving productivity (without removing people) and therefore provide social responsibility. This might not be the easiest path, but this is why leaders are paid more and granted more power. So take on this social responsibility and enjoy the higher reward levels of people thanking you for your leadership and trusting to follow your leadership

Your assumptions will change when transforming from traditional management to today's leadership style; in the past your assumptions of a subordinate might have been that they are lazy and unmotivated, but with the new leadership style your assumption will be that everyone inherently wants to do a good job.

There is a definite transition management has to go through, and it can be either very painful or actually enjoyable, depending on the understanding and motivation. What specifically is this transition? It has several areas: transfer of perceived ownership to where the ownership belongs, from directive to supporting, transfer of knowledge (part of empowerment), knowledge gain, from assignment of tasks to enabling, and understanding their value as a leader.

Transferring ownership—This is probably the single most important change to happen. Without it we will continue to see lackluster results from the shop floor. Self-directed teams must feel that it is *their* business, not management's. They work there, it is their environment, and they own the improvements and the failures. As an analogy, when your car breaks down, who decides where you get support from? Who decides if the solution presented to you from a service station is of value or not? What would you do if the service station had the authority to install an expensive and ambiguous repair solution? In my analogy the owner of the car is the self-directed team, and the service station that has authority to provide unsolicited repairs is the supervisor. Management always feels that they can provide a better solution for the process than the actual owners of the process, the self-directed teams. They probably can, *but* the solution typically fails because it is not owned by the self-directed team and can be too difficult to understand. Because of this, implementation of management's solution is by compliance, and slowly the solution is dismantled until they are back to the original state.

From directive to supporting is one of the later stages in the management transition process. The leaders of the plant are sometimes the main roadblocks. There are many changes that have to occur in implementing world-class results. This is one of the main reasons it can be very difficult to have a successful implementation. Please do not be discouraged, as the rewards of success are tremendous! The day-to-day pressures of meeting the production requirements almost always will override any world-class implementation. Amazing, when you consider that leaders are the ones implementing the plant strategy of world class. However, this is reality, and we always have to have a reality check. To start the transition from being a directive leader to a supportive leader you must first accept humility. You must accept that everyone within the organization can contribute something toward meeting the organization's strategy. Everyone has a role to play. As a leader, one of your main roles is to be supportive of people's efforts and creativity toward meeting the strategy. Go to the Gemba, ask questions about their progress, about roadblocks preventing this progress, about needed resources, and when you see a lacking skill, provide it. Use and develop your Mentoring skills, not your power.

Transfer of Knowledge—Knowledge sharing as a leader will not only increase the pace of effective decision making, it will also increase the trust of the people reporting to the leader. These two benefits alone will quickly translate into a competitive edge for the organization. Trust within the organization is difficult to achieve but when the leader is the pioneer of transfer of knowledge the culture rapidly follows.

From Assignment of Tasks to Enabling—Using power to impose what a leader wants people to do slows the organization down. This happens because the people always being assigned the tasks will develop a culture of waiting to see what the leader wants to do. Creativity is suppressed and risk taking is not part of this culture. Enabling people by investing in developing their skills will also provide a competitive edge. People will be engaged, feel valued, and feel secure.

The Management Steering Team must be skilled in managing change. It is not inherent, and if you think it is, you are well on your way to failing. The Management Steering Team will need to invest some time in obtaining change management skills.

At this point I would like to tell you of a real example of leadership. We had some mature teams who had reached the performing stage of the traditional team life cycle. The plant manager was asked to speak at a yearly, state-sponsored event for manufacturing companies of world-class strategy. The plant manager was provided with only 15 minutes to talk about self-directed teams. He decided the best and most effective method of communication was to have self-directed team members address the subject personally, but this was not communicated to the meeting organizers. This meant that the selected team members would personally have to present the benefits of working in a team environment. We asked for volunteers, and there were only a few. The two team members who eventually became the presenters were very nervous when we were traveling to the Baltimore Aquarium Center, where the meeting was being held. Their anxiety was beyond high. Upon our arrival, we found that newspaper, television, and other media were there. There were also many people (over 1,000) representing manufacturing businesses in the state. The great turnout was because the state governor was there. None of these media paid any attention to us. We were then all called to the conference room, where there was an agenda of distinguished speakers presenting their methods of world class. They were all authors of books and well known within their field. The conference room was extraordinary—the walls were actually the aquarium, where you could see the various fish

Team Stages to High Performance

Forming Stage:

- Providing a team structure with a facilitator, leader, scribe for problem solving meetings, and time keeper for problem solving meetings
- Establish which team members can be assigned expert roles such as TWI-JI Trainer, TPM, Quality, 5S, and TWI-JM
- Start developing team operating norms. What are some of the rules and processes all team members must follow in order to be efficient and effective. For example how will decisions be made, how will suggestions be processed, and how will the team address required discipline.
- Establish when problem solving sessions will be held. Weekly meetings or planned Kaizen events
- Will the team member roles change and how often

Emergent Stage:

- Development of Levels of Empowerment, what decision making and tasks will the team be responsible for. What is the time line?
- Assigned Key Performance Indicators (KPI) from Management Steering Team. Part of the organization's strategy deployment. Visual management of these KPI's are reviewed and supported daily with the Management Steering Team.
- Problem solving has started with support from their Facilitator
- Internal supplier and customer connections are made and part of the improvement program.
- A "Rite of Passage" for new team members has been established

Ownership Stage:

- Established Levels of Empowerment are almost completed and mastered
- Team is less dependent on their Facilitator for problem solving
- Regular internal supplier and customer improvement meetings are being conducted

- Responsible for their process scope operating budget
- Owners of the Standard Work for their scope of the process

High Performance Stage:

- Team pride is evident through their enthusiasm and creativity to go beyond assigned KPI targets
- Standard Work is reviewed and improved on
- All levels of empowerment have been achieved
- Team Leader attends Management Steering Team meetings
- Completed improvement and successful projects have increased to 15 per year
- Improvements are sustained
- Continuous skill development to meet changing needs are owned by the team

gliding by. The conference room also included a very intimidating podium in the middle of the room. This might not be a problem for people used to presenting, but to our two selected team members it was a serious concern. After all of the authors presented, the Governor of Maryland called our plant manager to the stage. He came to the podium and said a few nervous words. He then announced that if you want to hear what it is really like to work in teams, then you have to hear it from two people who live and work it every day in their daily work lives. They are extraordinary people who helped transform their work team from the Forming Stage to the High-Performance Stage of today. From that introduction, a special thing happened—the anxiety left; you could see the pride in their eyes as they walked to the podium. From there they just talked and articulated the benefits of having the ownership of their process. When they completed their communication to the people in that conference room, there was a measureable moment of silence and then a thunderous standing ovation. It was unbelievable and emotional. It was so rewarding. When everyone left the conference room, all of the news media, lights, cameras, and people conducting interviews were surrounding our two team members.

If you were selected as a speaker for your organization's world-class success, would you consider sharing the spotlight to the degree our plant manager did? Without question, this is powerful leadership.

This leader also valued having fun within the workplace to break down organizational barriers. If people were going to work hard and achieve extraordinary results, then he felt it was the responsibility of the leadership to provide rewards and recognition for the commitment. We never knew when the announcement over the public-address system would come: "Everyone must stop work now and get on the bus!" This was always a pleasant surprise—the whole plant would be treated to bowling, an afternoon at Pimlico Race Course, or a cruise on a dinner boat on the Chesapeake Bay. These events were not like planned team-building events; the only requirement was to get on the bus and have fun. I steadfastly believe these events significantly contributed to tearing down the hierarchical barriers within the organization and the departmental barriers, bringing people unequivocally together. Yes, this meant the complete manufacturing process was shut down for an afternoon. Did this ever create a missed on-time delivery? The answer is an emphatic no! This is leadership: investing in people and being confident of the expected result.

TRANSFORMATIONAL LEADERSHIP—ELIMINATE "US AND THEM"

First let's take a look at the first part of the lyrics of "Us and Them," a song by Pink Floyd:

> Us, and them.
> And after all we're only ordinary men.
> Me, and you.
> God only knows it's not what we would choose to do.

We as leaders must realize we are "only ordinary men." We do have a position to provide transformation, but it is through "Me, and you". Of course the "Us and Them" in this chapter refers to the management and the workers, respectively. Most organizations have the culture of "us and them," presenting a lack of trust, creativity, and contribution.

Let's now take a look, from an "us and them" perspective, at what each group might be thinking during common business activities.

Activity 1: General Communication Meeting

Us: What is important to communicate, what do they need to know, and what can I not trust them with? How can I present the bad news in a more positive light? Should I present the bad news? What kind of responses and questions should I expect?

Them: Do we really have to attend? It is difficult to understand the information. Am I allowed to ask...? I would like to present what we would like to improve and how we could contribute. We do not understand the metrics being presented. This is difficult to believe. What is expected from us?

Activity 2: Management Gemba Walk

Us: What is the output? Are we on target? What are the quality levels? What are the issues preventing us from meeting our targets? I am concerned that I am the only one providing solutions.

Them: Are they coming today? Get ready; here they come. What happened to my suggestion from last week? What are we supposed to do about these issues? How do they know? We will try their suggestion, but we know it will not work.

Activity 3: Executive Leadership Visit from Corporate

Us: I must prepare an agenda to present the positive messages. I will lead the tour so they hear what is needed. We must prepare and rehearse the management presentations. They do not want to hear from anyone lower than the department managers.

Them: I would like to tell them what it is really like to work here. We have contributed a lot; I wonder if they know this? Don't worry; you will only see them for about 10 minutes. I wonder what they thought? What do they talk about? They have visited twice this month; are we closing?

Activity 4: Problem Solving Session

Us: They are not capable of understanding this. I will need to steer them toward a solution that will work. Why are most of them not saying anything? That will never work. Everyone seemed happy with the solutions.

Them: This is confusing. I do not understand the discussion. I have an idea; should I say something? Last time they ignored my suggestion, so let them solve it. We were trained in DMAIC but never seem to follow the process. Just tell us the solution and we can save a lot of time. We missed the last three scheduled problem-solving sessions; where are we? They really do not understand this, but we do, so let us solve it. If continuous improvement is important, why do we need overtime to work on it?

Activity 5: Quality Issue Stopping Production

Us: Increase the sampling. Let quality look at it. We must get production started again. I am concerned this happens too frequently. The customer will never see it; we should be able to continue.

Them: This will require more work. Why don't we just try… This happens too many times. They preach zero defects, but we will let this go; production is first. What happened to my suggestion about this same issue? They are responsible for resolving quality, not me.

Transformational leadership works toward the elimination of "us and them" through the following:

Emulation—an effort to develop equality and be a model of the expected culture

Trust—developed through accepting and learning from mistakes

People development—providing higher level skills and the best opportunity for everyone to contribute to the overall strategy

Individual value development—through the autonomy of empowerment, encouraging people to explore new ideas and make decisions, leading to satisfied customers

Strategy alignment—100% of the people clearly know the strategy and how to contribute

Develop your organization through transformational leadership and realize the rewards of an engaged workforce.

PARTNERSHIPS MUST BE WIN–WIN

When you think of the word *partnership*, what comes to mind? Is it a business partner, your spouse, or a close friend? How about the people who work for you? In all cases a partnership must be win–win to be successful and sustained. This is not what the modern leader of most organizations seems to practice. On the website www.dailyjobcuts.com there is an average of 10 layoff announcements per day! I have provided below some recent examples of leaders failing in creating partnerships with their people. Some people might argue that the root cause is the economy. What about the lack of leadership? Why are some organizations succeeding while others are not? The economy is common to all of these organizations—leadership is not. Leaders who think of their people as partners will not fail. Why? Because they together will ensure innovation to achieve revenue targets and lower costs. Yes, even lower costs, because employees as partners will dramatically improve productivity and quality. So how do your partners determine what is winning? Many leaders struggle with this question and will even invest in reference books on the subject. How about asking your partners? I have provided below the top three companies to work for in 2013. If you are a leader, use your skills to develop win–win partnerships with your people and realize the success of meeting business targets. Don't get to the point of removing your most valuable assets.

Recent Examples of Layoffs

DreamWorks Animation co-founder Jeffrey Katzenberg says 350 job losses are the "right thing for us today" after the fantasy adventure has lost $83 million.

Goldman Sachs Group Inc. has an annual job-cutting process, and its equities-trading business is bracing for bigger cuts than fixed-income trading.

United Technologies Corp. noted in its annual report that it expects to eliminate 3,000 employees from its overall workforce in 2013 and close 1.85 million square feet of building space.

Yahoo! CEO Marissa Mayer announced that all staffers must work in the office, engendering the question whether this is simply a bid to end slacking at home or a harbinger of layoffs.

(All layoff examples from http://www.dailyjobcuts.com)

Top Three Companies to Work for in 2013

Below are the top three companies to work for in 2013, according to CNNMoney.com. Notice how these companies are developing winning partnerships with their people.

1. Google, for the fourth year in a row. What makes it so great? Not just the 100,000 hours of subsidized massages it doled out in 2012. This year Google added three wellness centers and a seven-acre sports complex, which includes a roller hockey rink; courts for basketball, bocce, and shuffle ball; and horseshoe pits.
2. SAS. With two artists in residence on staff, the perk-friendly, privately held data analytics firm takes creativity seriously. One employee cites SAS's "creative anarchy" as conducive to innovation. New this year: an organic farm for SAS's four cafeterias.
3. CHG Healthcare Services, a medical staffing firm. Employees compete in talent shows, trivia contests, and activities like the competition Dress As Your Favorite President. Extra paid time off is given to sales teams that meet their goals. This year CHG has added two on-site health centers.

Add Value to the Organization by Developing People

The new industrial society is giving way to the one based on knowledge. The new divide in the world will be between those with the knowledge and those without. We must learn to be part of the knowledge-based world. With the best man and woman for the job. To survive, we had to be better organized and more efficient and competitive than the rest.

—Lee Kuan Yew, Prime Minster of Singapore

Add value to the organization by developing your people and partners. This is one of the principles of the Toyota Production System (TPS). Unfortunately, it is one of least implemented even since the development of this proven system in 1948. Today, when most organizations start their Lean journey they start with a tool implementation approach like 5S, TPM, Kaizen, or Cells. Because these are very good tools, the organization realizes some early results. However, for many reasons the momentum gets lost and even the practice and application of these tools start to fade.

What Toyota realized, very early, was the value of their people—including the people in the process, who are actually touching and adding value to the products and services—to sustain results. With this realization Toyota included programs to develop their people as a key element of their strategy. Why is this not realized by most organizations? I believe it is fear of the loss of command and control by the organization managers. W. Edwards Deming taught Japanese top management how to improve design, product quality, testing, and sales through various methods. Deming offered 14 key principles to managers for transforming business effectiveness. The points which lay out the fundamentals of Total Quality Management were first presented in his book *Out of the Crisis*. Although Deming does not use the term *Total Quality Management* in his book, it is credited with launching the Total Quality Management movement. I think it is worth listing these principles here (Deming 2000, 23–24):

1. Create constancy of purpose toward improvement of product and service, with the aim to become competitive, stay in business and to provide jobs.
2. Adopt the new philosophy. We are in a new economic age. Western management must awaken to the challenge, must learn their responsibilities, and take on leadership for change.
3. Cease dependence on inspection to achieve quality. Eliminate the need for massive inspection by building quality into the product in the first place.
4. End the practice of awarding business on the basis of a price tag. Instead, minimize total cost. Move towards a single supplier for any one item, on a long-term relationship of loyalty and trust.
5. Improve constantly and forever the system of production and service, to improve quality and productivity, and thus constantly decrease costs.
6. Institute training on the job.
7. Institute leadership (see Point 12 and Chapter 8 of *Out of the Crisis*). The aim of supervision should be to help people and machines and gadgets do a better job. Supervision of management is in need of overhaul, as well as supervision of production workers.
8. Drive out fear, so that everyone may work effectively for the company. (See Chapter 3 of *Out of the Crisis*.)
9. Break down barriers between departments. People in research, design, sales, and production must work as a team, in order to foresee

problems of production and usage that may be encountered with the product or service.

10. Eliminate slogans, exhortations, and targets for the work force asking for zero defects and new levels of productivity. Such exhortations only create adversarial relationships, as the bulk of the causes of low quality and low productivity belong to the system and thus lie beyond the power of the work force.

11. a. Eliminate work standards (quotas) on the factory floor. Substitute with leadership.

    b. Eliminate management by objective. Eliminate management by numbers and numerical goals. Instead substitute with leadership.

12. a. Remove barriers that rob the hourly worker of his right to pride of workmanship. The responsibility of supervisors must be changed from sheer numbers to quality.

    b. Remove barriers that rob people in management and in engineering of their right to pride of workmanship. This means, inter alia, abolishment of the annual or merit rating and of management by objectives (see Chapter 3 of *Out of the Crisis*).

13. Institute a vigorous program of education and self-improvement.

14. Put everybody in the company to work to accomplish the transformation. The transformation is everybody's job.

I believe that managers who do not put principle number 8 in effect will not be able to carry the Lean journey momentum to the development of their people. This is unfortunate; successful organizations have been able to drive out the fear and then the momentum not only continued but created a fast pace toward achieving world-class results.

Deming also developed the System of Profound Knowledge. A quote from his book *Out of the Crisis* explains this system well (Deming 2000, pp. 23–24):

> Once the individual understands the system of profound knowledge, he will apply its principles in every kind of relationship with other people. He will have a basis for judgment of his own decisions and for transformation of the organizations that he belongs to. The individual, once transformed, will:
>
> - Set an example;
> - Be a good listener, but will not compromise;
> - Continually teach other people; and

- Help people to pull away from their current practices and beliefs and move into the new philosophy without a feeling of guilt about the past.

This, as Deming describes, is the basis for transformation of the organization. If critical knowledge is shared only by the few then the organization will only have the opportunity for slow progress, not fast. The people who are actually transforming the product or service by touching and working with it every day hold key point knowledge, which can prevent errors, make it easier, and prevent safety issues. This is what defines a key point when using TWI (Training Within Industry) Job Instruction: a very powerful program owned by the people transforming the products and services moving through the process steps. What are missing though are developed skills, organizational support, and leadership to bring out these process key points. The organizational support is discussed in Chapter 3 with the component of Organization Structure for Sustainment.

Deming's principle 14, "Put everybody in the company to work to accomplish the transformation," states it well: The organization needs the support of everyone. During the Lean transformation I was part of, leadership within our organization did not realize this until well into their tenure. "I wish I would have known this 30 years ago," became the favorite quote of the president. What he was specifically referring to were the results coming from the self-directed teams.

To reach world class, an organization must have a philosophy of not only meeting customer expectations but exceeding customer expectations. To do this you must—

1. Give people what they need and want
2. Break out of old ways of doing things

People, when challenged and provided with the needed skills, will exceed customers' expectations.

MENTORSHIP

Learning is not attained by chance, it must be sought for with ardor and attended to with diligence.

—**Abigail Adams, 1780**

With the implementation of self-directed teams, mentorship becomes an ongoing relationship of learning and developing, resulting in individual value and security. It is a two-way learning process. The mentor will end up learning maybe even more than the person or team they are mentoring. It does not take a special person to provide mentorship; anyone can embark on this valuable and self-rewarding skill. Mentoring is an essential skill for leaders in order to provide fast-paced development of people and to earn their trust. It demands commitment and preparation with objectives. The preparation of the objectives comes from the Management Steering Team—what the organization wants to achieve and how. Part of the how is by using mentoring to support, teach, and promote the strategy.

One synonym for the word *mentor* is teacher. Managers as teachers are absolutely a necessity in generating a fast pace. Other leadership qualities like support, effective communication, honesty, passion, motivation, and empathy are also required. My contention, though, is that if leaders think of themselves as teachers, these other qualities will also be there. Think of an athletic coach or a school teacher you remember—their primary or single focus was to teach, but they also had these other leadership qualities. If all they did was stand at the front of the class or on the field and read a script verbatim, would it be memorable, and how would you recall them? I propose not as a teacher. John Harbaugh, coach of the Baltimore Ravens, who recently won the NFL Super Bowl XLVII, was asked what the most valuable characteristic of a coach is. His answer: a teacher. There is not anything more personally rewarding than being a mentor and witnessing the invoked change.

Effective mentors are like friends: they make the learning experience comfortable and even fun at times. When the learner is comfortable and trusting, the learning process is easy and fast. The teacher is teaching what is required for the learner to contribute to the single-focus strategy. The teacher is providing the knowledge required for everyone to contribute. The assumption for the learning organization is that everyone can contribute; and with everyone having an influence on the achievement of the single-focus strategy, the organization is fast paced. Mentoring alone makes a significant contribution to the success of the organization. Too many organizations decide on certain skills required for people, provide the training from either inside or outside the organization, and then expect effective execution. If an initial training is required, this provides only 10% to 20% of the learning. Think of a book you read or even a journal; how much information did you retain? Mentorship is the needed method and involvement to achieve effective execution of the knowledge.

So mentors must learn how to teach and how to put the learner at ease in the environment right away with no distractions. The mentor is prepared with factual and beneficial knowledge and, if required, resources like documentation and experts. Mentors must want to teach. Mentors must be clear on how the knowledge transfer is going to contribute toward achievement of the single-focus strategy.

The Management Steering Team provides the following knowledge transfer execution plan:

- What has to be taught
- Why it is taught
- Who will teach and who will learn
- When it will be completed and what indicators will tell us the knowledge has been transferred

What has to be taught is determined by the single-focus strategy, how each group within the organization contributes to a successful business result. The Management Steering Team reviews each group to understand where key knowledge needs to transferred. This knowledge could be a better understanding of using statistical process control (SPC), preventive maintenance skills, customer requirement knowledge, using TWI programs, developing and supporting teams, financial methods, and so forth.

Why it is taught has to be very clear for each manager as a Management Steering Team member. If the knowledge transfer is perceived by the learner as just another training session, then the teaching session is very much a waste. The mentor must teach to make the connection with the single-focus strategy; when people understand why, it will first be remembered and most importantly be put into use. In fact the teaching should start with affirming the objective.

Who will teach and who will learn is determined by the interested group. If for example all self-directed teams must learn some preventive maintenance skills, then the mentor should be either someone from the maintenance group or the team facilitator for the self-directed team.

When it will be completed is determined by the Management Steering Team, who provides a knowledge transfer matrix that includes who the mentors will be, the subjects to be trained on, which groups to be trained, and the planned dates for each event. This team manages this plan to completion. Metrics are then established—*the indicators that will tell us the knowledge has been transferred*—to determine the impact on the single-focus strategy. These are lower level metrics, which will show the effect of the new knowledge in practice. It could be as simple as a 5S score or number of maintenance interventions provided by the self-directed teams.

Anyone within the organization can be a mentor when they have knowledge to transfer to others that will enable another person or group to contribute to the success of the single-focus strategy. At a minimum, each member of the Management Steering Team must become a mentor, because they are looked at as the leaders of the organization. They must understand they are not only teachers but also mentors. This means they must provide support of what they teach by being available for people struggling with the new knowledge. Retained knowledge from a classroom teaching is only about 20%. The knowledge transfer takes place only during practice, and for people to start practicing they must have confidence, which comes from the support and understanding provided with mentoring.

Providing only positive feedback is surprisingly not the only way to make certain the learner learns. Negative feedback can be explanatory in what has to be learned. It can show the learner where they have to spend more time in order to use the new knowledge more effectively. If both positive and negative feedback are executed correctly, it is very motivating for the learner. If either is executed poorly, it can be unmotivating, depending on the level of expertise of the learner. For example, a person with more experience with the new knowledge would rather receive criticism on where they need to improve further, while a learner just receiving the knowledge but with very little practice would rather receive the positive, "Here's what you did well."

People do not come preprogrammed with your newly developed single-focus strategy; so supporting, teaching, and promoting (a technique I call STP) it into place is critical. The process of mentoring can be completed anywhere, but the most effective place is in the Gemba. Use the tiered Gemba walk process by taking your direct report with you. Leaders are responsible for developing current or future managers. The application of the tiered Gemba walk will provide the environment for developing not only managers but also the self-directed teams.

It would be highly advisable for the Management Steering Team members to receive mentoring training. This was definitely one of the success factors in reaching our business results. Finally, a secondary benefit realized from the mentoring training was the motivation gained from the Management Steering Team members; they loved the training. It provided them with confidence. We all learned together, which was extremely rewarding. The Management Steering Team must understand they are mentors; they are the models, so they must develop themselves at a faster pace than other parts of the organization.

Recently in the news were comments from Dr. Connie Mariano about New Jersey Governor Chris Christie's personal weight issue. These comments were wrong and served no purpose. Why? Because she has a position of credibility, and like a manager who can easily criticize someone else's decision that did not end in positive results, the long-term effect of this criticism will always result in an environment of minimized trust. Dr. Mariano could have privately contacted the governor with positive and professional advice; she could have become a mentor; she had the opportunity. We as leaders must understand the impact of our statements, especially in the Gemba. Like it or not, we are looked at as mentors, so please do not make this same mistake.

CUSTOMER ORIENTATION

First, a definition of *customer orientation*: A group of actions taken by a business to support its sales and service staff in considering client needs and satisfaction their major priorities (BusinessDictionary.com). Business strategies that tend to reflect a customer orientation might include developing a quality product appreciated by consumers; responding promptly and respectfully to consumer complaints and queries; and ensuring a system to base decisions on the voice of the customer.

Supporting Sales and Service Staff

Three methods, described in more detail below, can be used to provide the required support of customer requirements:

1. Form an operational team and customer teams to jointly work on process improvements.
2. Effectively and directly communicate customer requirements throughout the value streams.
3. Form an operational team and a sales team to ensure service and quality improvements.

Form an Operational Team and Customer Teams to Jointly Work on Process Improvements

Forming these teams usually starts with some sort of crisis that happens to a key customer. First, you should not wait for this crisis to happen! There are too many benefits for both organizations to not start this key improvement program. I once worked for an organization that was definitely in a crisis with a key customer. The interesting part of this crisis was that the customer, who was at a higher level, did not want to form the process improvement teams. She felt it was too intrusive to her organization! In the end, a lower level manager realized the potential benefits and convinced the higher ranking person to go ahead with the program. We formed an operational team on our side and they formed an operational team on their side. We started off by determining the structure of the meetings:

- Who will lead the meetings: customer or us? We actually took this role.
- When will the meetings take place and where? We decided once every two weeks and we would alternate between their office and ours. The customer was the main decision maker here.
- What was the scope of the meeting? We both agreed it would start by addressing the main issues of what caused the crisis but would eventually evolve to improving the entire value stream for delivery and quality. We did not discuss it at the time, but as these meetings evolved, the improvement scope actually helped the customer to improve some of their processes, which were direct inputs impacting delivery and quality.
- During the first few meetings we had our sales representative as a participant, but she stopped participating, I think mainly because she was confident that we were actually providing value for her customer.
- We decided we would have meeting agendas prepared for the next meeting at the end of each current meeting.

The crisis we faced was very technical and involved many unforeseen issues with a new system and product. However, with the two teams focused on it, we actually resolved it within two meetings, and the new product system turned out to be more efficient than we had anticipated. With the crisis out of the way, we started to build the meeting agendas on other issues the customer would like to see improvement on. We would always follow the problem-solving methodology of DMAIC, and even though the customer's organization had a Six Sigma structure of Master Black Belts reporting directly to each Business Unit executive vice president (EVP) and then a Black and Green Belt structure below them, the customers on our team had no idea about DMAIC. So we taught them as we moved along on the various improvement projects. We once even received a compliment from one of the customer's Master Black Belts that their people on this team knew more and practiced more of Six Sigma than their own structure.

To me the most interesting time was at the start of each problem-solving session. We of course would start with the Define stage, and at this stage we would always get the same comments from the members of the customer team: "I did not know that" and "I wish I would have known that then—we could have helped." This to me is why this type of program brings benefits. The customer does not know, and normally they will help in order to improve the *complete* value stream. Our team even started to get invited to other parts of their operational processes to see where we could help. We once went on a benchmarking trip together to learn more about what Lean Sigma could do for us. Many times we celebrated together by going to a pizza shop or something insignificant like that, but we always had fun and it really contributed to our momentum. Finally, the improvements we made for the customer became world-class measures and their people would mention our program at various industry forums. Of course we also enjoyed process improvements, which helped our organization not just locally but globally.

Effectively and Directly Communicate Customer Requirements throughout the Value Streams

I propose that if your organization does not have an Organization Structure for Sustainment (component 3) of self-directed teams, TWI structure, new product introduction diverse team structure, and Lean Sigma structure, then your organization is very limited in basing your decisions on your customer. The value chain, at a macro level (see Figure 1.2), is not complex; but you cannot have broken links along the flow path of product and information. For example, in most organizations each and every person touching the product in manufacturing does not base their daily decisions on what is best for the customer. When I say this to manufacturing people, they argue, of course we do; we have an extensive quality system, we are

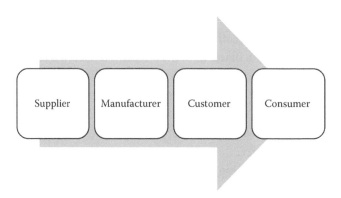

FIGURE 1.2
Value chain. (From Robert Baird.)

ISO certified, we are close to Six Sigma levels, and we spend significant time on training. All these are good arguments; but go to the Gemba and ask the workers about specifications, about what caused the last customer complaint, and about their interpretation of a defect, especially attribute. The outcome in the Gemba surprises managers. Why? Because they are not in the culture of tiered Gemba walks. I must also say this is not the fault of the workers; it is the direct responsibility of the management to ensure there is an effective system for 100%, yes 100%, of the people to be able to base their daily decisions on the customer. So the voice of the customer (VOC) must resonate throughout the value stream.

To enable this you have to have the organization structure I mentioned. The self-directed teams are connected directly to the external customer through the following:

- There is an internal customer system throughout the value stream. Each team is responsible for and accountable for their part of the process, and they are both an internal supplier and customer to other parts of the process. They have regular meetings with their internal suppliers and customers to resolve any quality and delivery issues. The specifications of quality and delivery are directly based on what is needed to ensure the external customer receives the best quality and on-time performance. The complete value stream is interconnected and interdependent with this system.
- Customer complaints are not resolved by the quality department or even with the add-on of an ad hoc process team. If a customer

complaint is received, the quality department determines which part of the process it came from and issues it to the responsible self-directed team. The team is empowered to determine the solution through a scientific problem-solving method, generate a report, which is reviewed by management and then sent to the customer. We also had the team members directly involved in any discussions with the customer. Finally, the team was responsible to visually manage the metric they used to determine the effect of the solution for another three months after the solution was in place. This ensured the solution was correct, and it also provided a lasting memory of this issue, so if this issue did appear again, they could directly base their decision on the customer.

- The daily management tiered Gemba walk provides effective communication about customer needs; using the techniques of STP they are able to support, teach, and promote the customer needs. Of course during the Gemba walk they are performing standard work for managers by executing their checklists aligned with the strategy. This provides another important connection throughout the organization.
- The organization structure also has a diverse team executing new product introduction (NPI) and they use the identify, design, optimize, verify (IDOV) process to ensure quality is built in and cycle time of this process is kept low. However, the most important element of this process is having the customer involved from conception to the first piece delivered. I cannot express enough the importance of having the customer directly involved. The customer here is not the sales person; it is actually the customer who is paying for and receiving the product for their organizational profit. So at the beginning the NPI team is bringing prototypes to the customer for direct feedback on the features and capabilities. Once it gets closer to a manufacturable product, the customer can again provide further comments and maybe suggestions for improvement. Finally, the customer and the NPI team review the first products off the assembly line.

By involving the customer in NPI, as I am strongly suggesting, the organization is definitely basing their decisions on the customer. Once again it also benefits your organization in ensuring quality (less cost) and NPI cycle time (faster to market).

The other parts of the organization structure like TWI and Lean Sigma are required for fast-paced execution, support, and learning.

Form an Operational Team and Sales Team to Ensure Service and Quality Improvements

It is not the main objective of these two teams to discuss the orders in progress and what is forecast—that is another meeting. The main objective here is prevention of possible service or quality issues. The two teams accomplish this by first narrowing the focus to two or three key customers and then going through this customer-oriented process:

1. Reviewing the two key metrics of on-time and quality data isolated to each customer. This review is looking for any possible trends that must be addressed before the customer becomes unhappy. There is also a review of the types of defects that are being pulled out through inspection for these customer products. You must assume that if inspection is finding them, some are escaping to the customer, so action must be taken. Some representatives of the self-directed teams are involved in these meetings, providing another connection from the key customers to the worker's decision making.
2. Reviewing past customer complaints: Have they been effectively eradicated or controlled? Are there smaller issues the customer has not officially complained about but it is now the second time the customer has mentioned the issue? Again, with the participation of the self-directed team members these smaller issues will have an opportunity to be stopped before they result in an official customer complaint.
3. Any previous actions taken from the two steps above are also reviewed and supported.
4. If operational teams are formed with the customer, then there is also a review of the progress or any roadblocks.
5. At various intervals of the above process, invite your customer to participate. This is to check if you are on a path of value for the customer.

As mentioned, in the beginning the focus is narrowed to two or three key customers. This is similar to the concept of a single-focus strategy, where in order to be fast paced in achieving results the organization must have their best and the most resources addressing this customer focus. The learning from this focused process will spill over to other customers. Because it is fast paced, once the process is functioning well, then you can start adding other customers to the program. This is what happens with a single focus: The result is not only a fast pace but also fast-paced learning, and finally a culture of how you are customer orientated is established.

MANAGEMENT STEERING TEAM

Have you ever assigned a key project, established a project leader to execute it during a busy time, and a few weeks later realized there was no support, no follow-up, and many roadblocks were in the way of success? So the project leader struggles throughout the year, completing other job responsibilities, and less and less of their time is dedicated to this project. Finally, it is the end of the year and there are very few results, so the project leader puts together a presentation of the process that was followed and presents the meager results along with an interpretation of why the shortfalls. This story of a poorly executed strategy is very familiar in many organizations.

Within the Six Sigma project world, why is Six Sigma belt certification rate after initial training less than 30% worldwide? And why do less than 20% of certified belts continue to do key projects? These can be attributed to lack of leadership support—talk about sunk costs!

Successful strategy deployment is not because of a quality formulated strategy but because of an effectively executed strategy. Too many organizations highlight a certain time of the year to formulate the strategy, spend a considerable amount of time on the formulation, and use very expensive human resources for this activity. Also, more than likely, the travel and entertainment budget receives a huge hit. After the strategy is formulated, it is cascaded down and most of the follow-up is through reports and monthly or even quarterly calls. I jest when I say, what an involved and then effective method for strategy deployment!

One key factor we fail to realize when studying the success of the Toyota Production System is that Toyota designed an organizational structure to support initiatives for improvement. They had their Senseis, visual management, Gemba walks, Kaizen teams, TWI structure, follow-up of project results to ensure sustainability, and other support. These provided fast-paced improvements because of daily support from management, knowledge sharing, systems that easily detected problems, accountability, and successful strategy deployment.

The above stated issues of meager project results, sunken costs of Six Sigma training, ineffective strategy deployment, and lack of support occur because of a lack of collaborated effort from the organization's management. The Management Steering Team is key in developing a fast-paced improvement culture. This team has the overall objective of strategy deployment and governing the organization. Because of these high-level objectives, the members are all the key department managers, and the top position of this team is the Management Steering Team Leader. This team meets at least weekly—monthly is not effective. Standard Work for Managers is used as the discipline for developing and ensuring momentum. The Management Steering Team is designed for effective and fast-paced implementation of the single-focus strategy. The time lines for implementation should always be aggressive. It should also be mentioned that the team meetings must practice effective meeting methods. Back in the 1990s, managers were trained on these techniques, but today we have lost these techniques, resulting in a lot of wasted time in meetings. In fact, ineffective meetings create more and unnecessary meetings. Here is a great resource for effective meeting techniques: http://www.effectivemeetings.com/.

All of the Management Steering Team members must fully realize that without daily Gemba walks guided by Standard Work for Managers and visual management, they would not be effective or fast paced. The business benefits they realize are as follows:

- Fast-paced strategy deployment.
- Key project completion times are reduced by 50%. Because of this they have more resources to lead and complete other required projects.
- Key project results exceed targets.
- A culture of engagement and creativity is created.
- Unknown superstars are realized.
- Department silos are naturally removed.

Why is the culture fast paced? Here are the key elements:

- Department silos are removed because the department managers are team members and they naturally realize during the discussions at the Management Steering Team meeting that other managers face some of the same problems. Managers are no longer rated on their department achievements but on how well they collaborate and on the success of the organization, and they witness other managers willing to help with knowledge and resources.
- There is a process they follow to formulate the single-focus strategy.
- The method of strategy deployment is well supported by all managers. It is their highest priority. This method is the daily tiered Gemba walk using Standard Work for Managers and STP techniques.
- Visual management is in place prior to starting the tiered Gemba walk. It allows for a fast pace because the displays (visual

Key Inputs	Process	Output
Leadership *(Management Steering Team)*		Quality
People		Delivery
Technology	Product or Service Process	Customer satisfaction
Machines		Environmental
Capital		Community
Materials		Costs

FIGURE 1.3
Leadership as a key input. (From Robert Baird.)

management) are at every point of the value stream and are owned by the self-directed teams. The visual management provides daily updates on the pitch—process output achieved for a certain time interval, usually every hour, the problems uncovered, the current project status, and standard work being developed and improved. It also provides daily updates related to the single-focus strategy.

- The self-directed teams cover improvements across all of the value streams.

With this approach the Management Steering Team members can make key adjustments daily, not monthly or quarterly. Workers become engaged and instead of 10% or 20% of the people directly involved in executing the single-focus strategy, now more than 80% are involved, with the goal being 100%. By the way, this 100% is achievable!

Considering leadership as an input (Figure 1.3) to your value stream provides a paradigm, a different view, on how you approach the business. If you weighted the importance of process key inputs, leadership would be heavily weighted. We must develop this input to world-class levels, just like the other inputs.

Management Steering Team Members

The Management Steering Team members consist of the plant manager as the leader, all department managers, human resources manager, finance manager, logistics manager, quality manager, and Lean Sigma leader; some steering teams also include a Shop Floor Team facilitator. You can add other positions, but this should be the minimum. As the team matures and learns to be effective in using decisions by consensus, the team should grow in membership and some members from lower levels in the organization should rotate. This will provide a more direct understanding of how the organization is governed and another direct communication method of the single-focus strategy.

Management Steering Team Implementation

The formation of the Management Steering Team is the first step in developing the components of a fast-paced organization. It is started by the organization's leader, and membership is the above-mentioned key department managers. It is not started by the organization's leader and then passed down for the Lean Sigma manager to provide the leadership; this will only result in either ineffective progress of the required fast-paced components or complete failure. Remember, the Management Steering Team is used for the organization's strategy deployment and for governing the organization—it is not for implementing Lean Sigma. The tools and methodology will be used in assisting and facilitating strategy deployment, but it is not the objective of the team to become experts in Lean Sigma. The first three implementation steps are critical, as this will be a major shift in how the organization is governed and the start of a new continuous improvement culture.

Implementation Steps

1. Implementation starts with the leader of the organization determining the objectives of the Management Steering Team:
   a. Formulate and execute the single-focus strategy.
   b. Develop the people to a higher level than the competition's.
   c. Break down silos; everyone's objectives and personal monetary incentives are based on the single-focus strategy.
2. The leader then determines the core members of the team:
   a. The leader's direct reports—in the beginning this must be limited to a maximum of six.
   b. A team facilitator working directly in one of the value streams.

3. The leader then prepares the communication to the members:
   a. Emotions such as feeling loss of power and fear of job loss or demotion must be addressed.
   b. Simplicity—do not make it complicated.
   c. Objectives.
   d. Working together.
4. First meeting request is called by the leader:
   a. Location of first meeting can be off site to show importance.
   b. Agenda is circulated with agenda items from the first three steps, listed above.
5. At the first meeting:
   a. Agreement of the new direction is discussed.
   b. Consensus is arrived at for the new direction.
   c. Symbolic document with the Management Steering Team objectives stated is signed by everyone.
6. At the second meeting:
   a. Description of new culture is defined.
   b. Formulation of the new single-focus strategy is started.

TIERED GEMBA WALK

When organizations go through their Lean journey, they start to put their own signature on the Gemba walk. Some managers go in groups, some organizations use only managers who are directly related to operations (HR, Finance, Engineering, Purchasing, Customer Service, IT, and others are not included), and various methods are used to communicate and implement the required actions. Also, the frequency of the walk varies by position. Organizations will even change the name of the Gemba walk: Shop Floor Management, Standard Work for Managers or Leaders, Walkabout, and the old Management by Walking Around.

There are three key objectives that must be satisfied during the Gemba walk:

1. Business results are achieved and sustained at a faster pace than the competitors
2. All stakeholders realize the value the Gemba walk brings
3. Support, teach, and promote (STP) the single-focus strategy

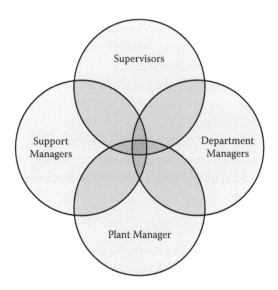

FIGURE 1.4
Tiered Gemba walk relationship. (From Robert Baird.)

I like the tiered Gemba walk approach, where each manager, starting with the Shop Floor Team leader, goes on the Gemba walk with their direct report (see Figure 1.4). So it is tiered like this:

- Team leader starts the day by reviewing the visually managed KPIs with their self-directed team and organizes the work to be performed.
- The team facilitator (the transformed supervisor) walks to each of the team leaders' cell visual management boards to provide support of production and continuous improvement efforts related to the single-focus strategy.
- The department manager walks with the team facilitator ensuring the checklists (Standard Work for Managers) have been completed and can help remove any roadblocks toward the day's productivity and continuous improvement efforts.
- The plant manager walks with the department manager, again reviewing the checklists and providing support.

No Gemba walk should take longer than 15 minutes. If it does, then the visual management should be questioned: Is it really telling us what we need to know? So what do these Gemba walks review? This is determined by the Management Steering Team, who develops checklists for each management position. Each item on the checklists is a minimum review, check, or observation required to ensure fast-paced execution of the single-focus strategy. These checklists are the Standard Work for Managers performed every day. Another benefit of the tiered Gemba walk is that each manager can use the STP technique to develop their reports and make certain that everyone understands the single-focus strategy and everyone knows how they contribute. At the end of each Gemba walk the main findings and assigned actions are recorded on the Gemba board to communicate to interested parties.

For optimum results, the operation support managers like HR, Finance, Engineering, Purchasing, Customer Service, IT, and others must also participate. What do they review, check, and observe during their Gemba walks? Here are some examples: HR of course plays a significant part in any Lean journey. People must go through transformations in their job requirements and responsibilities. Further skills need to be obtained, like problem solving, using Lean Sigma tools, and working within a team organization. Once the Management Steering Team has developed a plan to transfer these skills to required job positions, the next step is to implement the plan. HR, who is on the Management Steering Team, now plays an important role and uses STP (support, teach, promote) during their Gemba walk. If the organization has a hierarchy of positions within the HR team, the tiered approach should be used. Engineering can check and determine methods of improvement for machine settings. Purchasing can observe the effect of the process on the raw materials used. Customer Service can get real-time updates of critical orders to provide feedback to customers on the time line of the process. Again, even for the support processes, *review, check, and observe* must be aligned with the single-focus strategy developed by the Management Steering Team. The more stakeholders aligned with the single-focus strategy, the faster the organization becomes in accomplishing what is required to delight customers.

Supporting departments can also be part of the daily Gemba walk by going along with the team facilitator, operational department manager, or plant manager if an issue was discovered on the previous day's walk. For example, one of the self-directed team members might have found an issue that was intermittent and was difficult to realize the root cause. The team leader might invite the Quality manager on the next Gemba walk to review at the point it is happening and to determine some corrective action. Also, the operation manager's Gemba walk can be designed to always include support managers.

As the organization's leader, what do you want to accomplish? Do your employees know what needs to be done to reach objectives? Do they know how you expect them to behave? And—once they know the *what* and *how*—do you provide them with enough autonomy to get the job done in an effective and timely way? The answers to these questions are communicated and supported during the leader's Gemba walk. In the tiered Gemba walk, the leader also takes the opportunity to ensure all of his or her direct reports are also communicating the answers to these questions to their direct reports. All management levels and departments must understand they have the responsibility of supporting, teaching, and promoting the single-focus strategy. With the strategy clear, there is no more guessing as to what the direction is, and therefore there are fewer wasted resources that need to be improved in areas that do not contribute significantly to the strategy. With fewer wasted resources, there are now more resources available to work on what matters to achieving the single-focus strategy, and therefore the organization is faster.

Effective Communication

Using effective communication methods is key. The tiered Gemba walk provides an effective, proven structure and method for supporting, teaching, and promoting your organization's strategy, culture, and fast-paced business results. We now need some rules for successful communication. Here are 10 proven rules:

1. Always try to give feedback based on facts and not on opinions or emotions, which might upset or offend the other person.
2. Always try to empathize. Try to accept the other person's views without preaching and/or moralizing.
3. Criticize using neutral language and tone of voice; then ask for their response.
4. Say what you mean without becoming sarcastic.
5. If you want something from others, ask, don't command.
6. Give the other person a chance to speak; storytelling does help, but keep it brief.
7. Always explain why something needs to happen; don't threaten.
8. Don't give advice or opinions if people don't ask for it.
9. Be to the point; avoid vagueness at all cost.
10. Don't talk down or up to others; avoid diverting the conversation to trivial matters.

These 10 proven rules and other communication methods discussed later in this chapter, in the section titled "Something about Communication," are key to generating the pace toward meeting the business results the organization is striving for. A word of warning: The tiered Gemba walk can be perceived by others as pompous, and the "us and them" attitude will only be strengthened. Management must be clear they are leaders there to support, teach, and promote; empathy is probably the single most important practice when executing the tiered Gemba walk. Yes, we must support, teach, and promote; but understanding how to do this effectively will provide the motivation and success.

Tiered Gemba Walk Implementation Steps

The tiered Gemba walk is designed by the Management Steering Team to align with, communicate, and execute the single-focus strategy. As you can see from the example tables in Figures 1.5–1.8 the checklist items are focused—not a long list as an audit might have. The Management Steering Team must design the observation and review toward things they know are key to achieve the single focus. Following are the steps to take to implement the tiered Gemba walk:

1. After the single-focus strategy has been established, each checklist, by job position, can be designed.
   a. Each item on the checklist is observed or checked during the daily Gemba walk. See Figures 1.5–1.8; these check sheets were designed for a single focus of quality.
   b. Figures 1.5 and 1.6 are for the job position of team facilitator/ supervisor. For explanation, the Scrap box review is for the defects the operator found that shift and to determine if there are any Type I or Type II errors. If so, this is a great opportunity to teach. The

	Observations	Improvement Notes
☐	Quality Control Check	
☐		
☐	5S Score	
☐		
☐	Machine Condition	
Complete each observation per week		

FIGURE 1.5
Supervisor checklist. (From Robert Baird.)

Daily Improvements	Improvement Notes
☐ TWI Matrix	
☐	
☐ Key Metrics	
☐	
☐ Team Project Progress	
☐	

FIGURE 1.6
Supervisor checklist for improvements. (From Robert Baird.)

TWI matrix is part of the TWI-JI program. It provides who will be trained, on which day, which job breakdown, by which TWI-JI trainer, and who will provide the observation of this trained person performing standard work. It is key in getting standard work into place.

c. The team facilitator should complete the checklists on the daily Gemba walk in the morning. The Gemba walk must take place at this same time every day.

d. The first checklist is for observations, and the second checklist is for reviewing the visually managed metrics at the cell team board with the team leader.

e. One observation is selected each day and performed with one of the team members at a cell. Findings are noted in the Improvement Notes section. All observation categories are completed by week's end.

f. The second table provides what has to be checked every day for progress and support. Again, findings are noted in the Improvement Notes section.

g. All improvement notes are transferred to the Gemba board for other stakeholders to review.

2. The team facilitator completes their daily Gemba walk with each of the team leaders from the self-directed teams at a certain time of day; the time of day selected is recorded in a program like Microsoft Outlook for the entire year.

3. The department manager (boss) and the team facilitator (direct report) take the Gemba walk together, again at a preselected time of day. The department manager reviews the two checklists with the team facilitator, discussing as they walk to the cell visual management boards. The second checklist's Daily Improvements are

Metric Board Review	Improvement Notes
☐ TWI Matrix	
☐	
☐ Key Metrics	
☐	
☐ Opr. Project Progress	
☐	
☐ Yield Levels	

Complete each review weekly with each of the Department Managers (use Support, Teach, Promote)
- *Ensure accountability for the results*
- *Ensure fast pace of key project completion*

FIGURE 1.7
Department manager checklist. (From Robert Baird.)

reviewed at each of the cell visual management boards, because data and information for this list is part of the cell's visual management.

4. Finally the plant manager uses their checklist (Figure 1.7) to go on their scheduled Gemba walk with the department manager. The review takes place again at each of the cell's visual management boards. The role of the plant manager is more in the application of STP, but obviously progress is reviewed.

5. As mentioned in the explanation of the tiered Gemba walk, support managers must also be part of this program. Figure 1.8 provides an example checklist. How does this align with the single focus of quality? The TWI training matrix is scheduled training to reduce variation; employee survey actions were from comments made about improving quality; the Career Path added skills related to identification of root causes; and a training program was used to provide Poka-Yoke knowledge.

6. All findings from the day are recorded on the Gemba board, along with any actions determined and who is accountable. These are typically

Employee Reviews	Improvement Notes
☐ TWI Matrix	
☐	
☐ Employee Survey Actions	
☐	
☐ Training Program Results	

Complete each review weekly with all shop floor positions (use STP - Support, Teach, and Promote)

FIGURE 1.8
Human resources checklist. (From Robert Baird.)

Leaders Standard Work Completion Board					
Manager	Monday	Tuesday	Wednesday	Thursday	Friday
George					
Mary					
Janette					
Jimmy					
Gary	███████				
Scott					
Daniel					
Katelyn					

FIGURE 1.9
Gemba walk board. Black box denotes green. (From Robert Baird.)

actions that can be completed within one day or only require further follow-up. The plant manager plays a key role to support the people in removing any roadblocks and promoting the entire program.

7. Figure 1.9 shows an example of a board that is used to track who completed their Gemba walk. The names of the managers are on a tag that is red on one side and green on the other. When they complete their Gemba walk for that day, they turn it from red to green. I do not necessarily promote this control, as the tiered Gemba walk should provide enough value for each of the managers to participate on their own motivation.

SOMETHING ABOUT CHANGE

It is arrogant to think that we can influence change in others without feeling the need to change something in ourselves as well. After all, change is a learning experience in itself. If we believe that it is for everyone but us, we are likely not asking the right questions or enough questions, or not paying attention to what is going on around us. Whatever types of change people encounter, certain patterns of response occur and reoccur. It is important that change leaders understand some of these patterns, since they are normal outcomes of the change process. Understanding them allows leaders to avoid overreacting to the behaviors of people who, at times, seem to be reacting in mysterious, nonadaptive ways.

There are six discernments of change that employees will show whenever they are exposed to change:

Employees will initially try to change, but it will feel uncomfortable. When you ask most people to do something differently, they will initially make an attempt to adapt. During the first communications to define what is expected, most of your employees will attempt to accept the change. This beginning communication step is the most sensitive time of any change effort. The message must not be ambiguous by trying to include too many messages. The main objectives are to define the change and the benefits to the business, and if the change is not done, what the consequences are. Managers must be prepared to take the next step and use the STP techniques. This preparation of using STP is planned by the Management Steering Team.

Some employees will reject change and some will accept. Some employees naturally love change; they always seem to immediately see the benefits. Others love stability, they are conservative, and when change is proposed, anxiety immediately goes up. Recognize the employees who more willingly accept change, and work with them to become change agents for the others. If you do not have a self-directed team structure, these people can provide the benefits of positive peer discussions.

Employees will become confused as to what is expected if you introduce too much change. Be careful not to introduce too much change. For example, if you are introducing a self-directed team organization, many specific steps are needed, including introduction of the reasons and the benefits, training, forming, skill development, empowerment, role transition, and others. Implementing all of these steps in a short period of time will create anxiety and confusion, the result being failure in the change. The Management Steering Team must have a plan to understand the signs of acceptance in each of the implementation steps. Inform employees of the plan for implementation and communicate frequently enough so that the employees have learned the plan. Think of yourself as a passenger in a plane sitting on a runway waiting for clearance to take off. If the pilot is providing updates frequently enough, your anxiety and patience will improve.

Employees will be more comfortable about the change if they feel part of a team. If you are already organized in a self-directed team environment, any change effort happens at a much faster pace. This is because employees have the benefit and comfort to discuss their understanding of the change with their peers. These discussions will accelerate the acceptance of change as they will not feel alone in going through the process of change.

Employees will be looking to see what management is going to provide. Employees will feel at ease and will accept the change easier if management demonstrates their support for the change. Too many times the leaders will communicate a need for change and then leave it to the workforce's meager resources to put it into place. For example, an organization announces that everyone will now be responsible for problem solving. They provide classroom training and assign some improvement projects to various groups. The managers then go back to managing their traditional tasks and back to command and control. Management must understand that the communication step is the first step. The next step of practicing the techniques of STP during their tiered Gemba walks is the critical step of making the change stick. This alone will demonstrate to the workforce that we are in this together.

Employees will look for crimps in the armor. Unfortunately, even the managers might not be ready or willing to accept the proposed change. This can be catastrophic, as they will secretly start supporting the people who are skeptical about the change, the part of the workforce you know are critical for the change. The design of the tiered Gemba walk will provide assistance in nudging these managers toward becoming advocates of the change. Designing the Standard Work for Managers checklists will also help in supporting the change.

CONCLUSION

It is important for leaders to anticipate and respond to employee concerns and feelings, whether they are expressed in terms of practical issues or emotional responses. When planning for and anticipating change, include a detailed reaction analysis. Try to identify the kinds of reactions and questions that employees will have, and prepare your responses. Above all, change requires leaders to work on making it happen, daily. Practice the techniques of STP and use Standard Work for Managers on a daily basis. Remember that the success of any change rests with the ability of the leaders to address both the emotional and practical issues, in that order.

Each of the managers must understand that change does not happen after the initial plantwide communication of the required change and the compelling reason. Change happens when we "walk the talk" and start communication on a daily basis. This must be consistent, and the message must always be related to the reasons for change and how we can support it. Each manager must have empathy when they are on the shop floor—understand that mistakes will happen. They must understand that mistakes are probably related to the environment and not so much to motivation.

Another powerful group is the shop floor self-directed teams. First, this structure must have already been established. Once established, these teams must be involved in activities and decisions related to the desired change. They are the largest group of people in your population, so leaving them out will obviously not work and will slow down the process of change. Participating in self-directed teams will take away employees' sense of being alone through this change. Team participation will also provide the required ownership to help support the change.

Something about Communication

I hear and I forget. I see and I remember. I do and I understand.

—Confucius

One-Way Communication

No news is *not* good news. If you are not receiving any feedback on what you are attempting to communicate or change, you can be assured that it is one-way communication and the communication effort was a significant waste. Of course, this is very ineffective and should be considered a failure of the communication effort. Do you have a communication system where you post beautiful charts and diagrams of information you would like communicated? How much time and effort have gone into these postings? In your next meeting with whatever groups of people (managers, supervisors, teams, etc.) ask them questions about the information posted. I believe you will be surprised; this is an example of one-way communication. Beautiful charts and diagrams are assumed to be understood by the people transmitting the message. For some levels of the organization, the fundamentals of charts might not be understood. For management, the receiver of information in this case, the information posted might not be high on their priority list of information to be used and understood. We actually had a camera on the area (for other reasons) where most of the Key Performance Indicators were posted, and the video revealed that in a four-hour period only one person stopped very briefly to look at these posted measurements. The result is a wasted communication effort and ultimately a lost tool for improvement.

Communication within an organization can be pictured as individual bubbles that are only sometimes impregnated. Management has one language, Engineering another, Finance another, and the shop floor yet another. These departments all have their own paradigms, which can prevent the messages from being accepted or cause significant slowing in effectiveness. For example a Democrat's opinion on gun control is almost impossible for a Republican to understand.

From another perspective, adequate communication may be perceived as successfully completing the objective. Thus for passenger air transport, the very formal and consistent communication process used by pilots, co-pilots, and air traffic control may be considered adequate if the passengers were delivered to their expected destination at their expected time. Would this be a model of effective communication and could it be replicated for all organizational communication? We must understand how to communicate to other departments and especially the shop floor, where it is always difficult to achieve clear transmission of objectives. This is mainly because of the difference in culture between management and the shop floor, the "us and them." However, once the managers understand this difference, they can work toward methods for impregnating these bubbles.

For example, we gave a shop floor self-directed team the objective of redesigning the performance appraisal to meet the objective of developing to a high-performance team. The current and traditional appraisal system has an outsider (management) paradigm with categories such as communication up and down, team player, diversity, and planning. It is also an annual event. The performance appraisal program designed by the self-directed team had a much-reduced criteria list of 10 items, which included practical requirements such as attendance, getting along with others, completing team-assigned tasks on time, knowledge of operations, helping the team on the floor, and so forth, and they appraised each other quarterly. This made more sense and was the team's ideas of value added as a team member—and best of all it worked toward achieving higher motivation and productivity. You can question and debate all you want about these practical categories, but the result spoke for itself.

Recently the federal government has realized the importance of communication in the language of their customers. In the past, tax notices of nonpayment were ignored by many small businesses. The notice was typically two pages in length and contained legal terms not well understood by most small business owners. The solution was to reword the document and shorten it to no longer than a paragraph. The result was 100% compliance of the notice.

What can we do? Whatever information you feel is critical to the success of your business, you need to have empathy and find communication media that are effective. When you feel you have an important message to communicate to the shop floor, do you make the mistake of adding other messages since you have their attention? After the delivery of the messages, what was the Q&A like? Did people start asking questions related to everything but the message you were trying to communicate? Which messages did they leave the meeting with, theirs or yours? In the book titled *Made to Stick*, authors Chip Heath and Dan Heath (2008) emphasized six key qualities of communicating a message: simplicity, unexpectedness, concreteness, credibility, emotions, and stories. With simplicity the rule is to communicate one simple message to make it stick in the minds of the audience.

We used a communication method we labeled Nine-Minute Meeting. Here are the rules:

- The communication cannot be longer than nine minutes.
- There is only one message.
- The meeting must take place on the shop floor or where the people work.
- The communication can consist of only one message.
- There is no opportunity for question and answer after the message is communicated.

There is a shop floor language we have to understand and respect. The language difference is mainly related to levels of understanding and motivation. These two differences, if not understood and resolved, will provide a serious roadblock to world-class results. Leaders within the organization must understand that this is their new role: communication and being a model of world-class methods. Being a model is a powerful way of communicating. People watch how leaders react to certain situations. If leaders react by fire fighting and not addressing the system as to why a failure occurred, credibility is lost—and credibility is key in communication.

With general communication, repetition is required. The leaders of the business need to use the critical information in every communication forum they are involved in: monthly meetings, Gemba walks, operations meetings, reward functions, company newsletters, websites, one-on-one praise, and the Christmas party. This means every meeting (one method of communication) starts with the single-focus message. Repetition of the information is key. We have used a PowerPoint presentation with various points of information being projected on three different walls within the plant. This was one medium that helped in our communication of critical information. How did we know it was helpful? We received feedback from almost every level of our organization! There were suggestions for more information and suggestions on how to improve the objectives we were broadcasting.

Another example of a great exercise to improve the total communication of your effort is to follow these steps:

1. Provide a list for each of the following;
   - What are the important messages you would like to be communicated? (No more than two at a time, one preferably.)
   - Who are the groups you want the message to be understood by?
   - What are the noise (distraction) factors?
   - What are your current methods or media for communicating your world-class efforts?
   - What are some emotions we will be addressing?
2. Select the top one or two messages; you would like to make sure it sticks. Consider the receivers' emotional response.
3. Provide the internal and external noise factors that might interfere with the communication of your message(s).
4. Provide a list of the groups you would like to deliver your messages to.
5. From your list of current methods and media of communication, determine the best for each of the groups. You want full understanding of your top two messages.
6. Improve on the methods and media required to deliver your messages for each group.
7. Determine feedback methods to determine if your communication was successful.

Example

Step 1—Important message list:
- New style of management, changing from traditional to coaching
- World-class model

- Bottom-up communication
- Information systems provided for each cell
- Cell teams have production and quality responsibility
- TPM concept of error-proofing machines with work team participation
- Performance indicators
- Strategic plan

Step 2—Top two messages:
- World-class model
- Strategic plan

Step 3—List of current methods/media of communication:
- Gemba walk
- Company newsletter
- Employee survey feedback process
- Bulletin boards
- Team leader review meetings
- Team facilitator meetings
- Training sessions
- Customer plant tours
- Information meetings
- e-Mail

Step 4—List of noise factors:
- Recent layoff
- Conflicting messages from management
- Sales staff location
- Corporate-level plan not fully committed to lean

In the communication process there is a sender (the person with the message) and a receiver (the person receiving the message). In the traditional explanation of this communication process there are a number of things that the sender and receiver can do to facilitate the intended communication.

When humans are attempting to communicate, the following elements are as such:

- Information source—the source of where the intended communication begins. This can be a number of media, such as books, Internet, television, radio, and the human brain.
- Transmitter—the person attempting to translate the information to the receiver.
- Channel—the method of transmitting the message, such as audiotape, PowerPoint, video, e-mail, telephone, web meeting, etc.
- Receiver—the person or group of people the message is directed at.
- Destination—the form the message ends up in. This could be memory, notes, document, book, tape, etc.

EFFECTIVE COMMUNICATION WITHIN ORGANIZATIONS—INFLUENCED BY THE STRUCTURE

To effectively communicate, we must realize that we are all different in the way we perceive the world and use this understanding as a guide to our communication with others.

—Tony Robbins

Communication that happens within organizations can be very complex, with many messages coming from many directions. We all know communication is absolutely essential to organizations. We also know that the more times a crucial message is passed from one group to another, the message gets distorted or even completely transformed. Different people interpret the exact same message very differently. Communication happens in many forms: email, meetings, speeches, conference calls, videoconference, presentations, written memos, and other more casual methods like discussions at the coffee machine or over lunch. These are most of the common methods most organizations communicate. But how effective is the communication? What happens to the message when it travels throughout the organizational structure? Does it get to the right people at the right time with the right methods? If it does not, then important messages are ineffective and the organization suffers. Because of ineffective communication some people will work on the wrong things, strategy deployment and policies can be significantly slowed, and more resources are required. Effective communication will increase the speed of communication, and the organization will use fewer resources, becoming lean and customer oriented at a faster pace than their competitors.

The noise factor is at its peak when management is attempting to transmit important messages such as the yearly business strategy; the company's vision, mission, and values; policies; and customer requirements. The noise is created because management is too far away from the message recipients. These important messages are transmitted through many levels of management, but even one level is too much. Each message should be validated one management level at a time to ensure clarity and understanding, and the transmission cannot go any further until it is validated. As you can tell, this communication method takes much too long and is thus ineffective.

To achieve effective communication, securing the many benefits, organizational structure must be the focus for improvement. The traditional hierarchal structure ominously impedes effective communication and with it the organization functions at a slower pace, creating miscalculations of the intended strategy. With the many layers of a traditional hierarchical organization, the key messages of strategy and policies travel a torturous path of departmental biases and paradigms. Each layer of management will put a spin on the message to best suit their interest; it is how they have survived in the past. Even if they communicate these key messages verbatim, people's departmental paradigms will translate the message toward their past tribal success. Each department will contribute, but the global contribution toward the objective will take longer and the achievement of the target is most likely less than what is possible. This does not mean they might not meet the intended target; it means the possibility of meeting the full potential (most likely exceeding the target) of the organization is undermined.

The solution is to flatten the organization with a team structure. Job titles are not really important in an organizational structure of teams. Job titles actually get in the way with emotional responses. What is important is how the organization communicates and the methods of governing. With the organization being structured in management teams and value stream teams, the organizational structure essentially has only three layers: executive team, functional teams, and value stream teams. The executive team governs and sets the strategy and policies of the organization; the functional teams determine the required resources, technologies, and marketing; and value stream teams execute according to plans. The key messages now have to travel from the executive team to the functional teams to the value stream teams, a much less tortuous path than the many management layers of a traditional hierarchical organization. The team structure will remove the paradigms because of the speed of communication. The speed will also remove the distortion of the intended key messages. Fewer resources will be required to execute the strategy because the objectives of the organization are clear, enabling people to work together. The most short-handed goals ever scored in one National Hockey League game by one team occurred on April 7, 1995, when the Winnipeg Jets scored four. This was because, each time, the coach effectively communicated to the players (value stream team) the objective of scoring a goal and not defending. The paradigm is to defend, but effective communication allowed fewer resources to achieve extraordinary results. Imagine if a professional sports team had a traditional hierarchical organization like most companies today; execution speed would leave them noncompetitive. Most professional team owners, like Jerry Jones of the Dallas Cowboys, take a participative role because they realize the importance of effective communication through direct communication.

If you are not in a position to reorganize the complete organization, then do it locally. For example, if you are in one location in an organization of many locations, restructure the local organization into teams. The local leader and the direct reports are one team (I call this the Management Steering Team) with the responsibility of strategy deployment and governing. The people working directly in the value stream are structured into self-directed teams.

This might seem like a radical change in organizational structure, but I refer to a quote by Albert Einstein, on the definition of insanity, "Doing the same thing over and over again and expecting a different result." As an organization do not expect a different result from repeating traditional efforts; change the organizational structure to gain effective communication and achieve the business results your people are truly capable of.

Something about Bottom-Up Approach

It has been shown, and I have experienced also, that without a strong leader driving the Lean Sigma journey, the chances of receiving world-class results are very small. It has even a lesser chance of succeeding when management starts the Lean journey initiative, passes it on to a lower part of the organization, and then goes back to what they have always done. Why does it have a lesser chance of succeeding? Because people basically do what their leaders do. How many lone Lean practitioners have fallen victim to this? So if you have to start the Lean journey from the bottom up, here are some key points:

- Knowledge transfer—Start training sessions of Lean tools and methods. Always include how these tools apply within the process and what business metric they affect and improve. Bring in the popular Lean books and start a reference library in a central location. Get the support of the leader to read at least one of the books, like *The Toyota Way* (Liker, 2004).

- Talk the leader's language—Understand what matters to them, what drives them. You must understand the budget lines and what affects these lines. A leader will always have responsibility for the budget, so understand this and work this into your message.
- Go see it—Take the management team and others to other organizations that are well into their journey and have known success. Part of winning the Shingo Prize is that companies agree to hold tours of their organization. Get onto Lean forums and you will soon realize that organizations are more than willing to share their success; they are very proud of what they have accomplished.
- Hear about it from peers—There are CEOs or local operation managers that have already gone through their transformation and continue to be successful; they are totally immersed in Lean. You can find these people from your network, such as people you went to school with or your vendors, or at a Lean conference where these names are known.
- Do it—Go to the Gemba and start a pilot. Successful criteria for selecting the pilot area are as follows:
  1. Improving this area or process step will directly and positively impact the organization's KPIs.
  2. There are some low-hanging fruit.
  3. It is a key step in the process.
  4. The people within this part of the process accept new challenges and are willing to learn.

Start an A3 problem solving report. This tool provides critical project information, all on one page (A3 paper size), like the background of the problem, the current condition, the goal/targets, analysis, and recommendations. Use it to determine the target. Outline the plan and then present to management. With approval, start visual management of the key metrics and get the workers within the process involved by updating the measure, training them on problem solving, and allowing time to problem solve. The visual management at the process is very powerful, and when you have an opportunity to bring the leaders to the Gemba, show them your progress. Finally and most importantly, make this project your single focus; look for and promote progress every day. Give yourself a goal of achieving results within one month. The results must include a financial result, preferably a hard cost, something the leaders can see on the profit and loss statement. Set up a meeting to present the results to the management team. During this presentation you must clearly bring out the contribution of Lean Sigma in the achievement of these results. You can then ask for more of an organizationwide implementation.

REFERENCES

Booton, Jennifer. 2012. Manufacturing Renaissance? Exports, Reshoring Could Bring 5M Jobs to U.S. *FOXBusiness* (September 21) http://www.foxbusiness.com/economy/2012/09/21/manufacturing-renaissance-exports-reshoring-could-bring-5m-jobs-to-us/#ixzz2cd9h05He

Business Dictionary.com. http://www.businessdictionary.com/definition/customer-orientation.html

CNNMoney.com. http://money.cnn.com/magazines/fortune/best-companies/.

Deming, W. Edwards. 2000. *Out of the Crisis*, pp. 23–24. Massachusetts Institute of Technology, by permission of The MIT Press, Cambridge, MA.

Ernst & Young, LLP. 1998. Measures that Matter. Boston. http://valuementors.com/pdf/Measures%20that%20Matter.pdf

Heath, Chip, and Dan Heath. 2008. *Made to Stick: Why Some Ideas Survive and Others Die.* New York: Random House.

Kiechel, Walter. 1982. "Corporate Strategists under Fire." *Fortune*, December 27. pp. 34–39.

Leading Through Connections: Insights from the Chief Executive Officer Study, IBM, 2012.

Liker, Jeffrey K. 2004. *The Toyota Way: 14 Management Principles from the World's Greatest Manufacturer.* New York: McGraw-Hill; 1st edition (December 17, 2003)

Pink Floyd. 2009. "Us and Them." Warner Music UK Limited.

2

Component 2: Process Design and Visual Value Streams

This is component 2 because it is the next part or step of fast pace which must be in place. The value stream flows must be clear and intuitive for everyone and even the new people just starting with the company. The process must be designed in such a way to be conducive to continuous flow and low work in progress (WIP). Lead time is a differentiator for your customers, and process design is the principal contributor. To understand where the organization is in terms of achievement pace, the value streams must be visual. This visual management will facilitate the support required by the leaders. For this component I discuss several different methods to having your products or services moving through process steps faster and with quality assurance.

MANUFACTURING AND OFFICE CELLS

A manufacturing or office cell is a configuration of process equipment (or office steps) designed to take advantage of similarities in how the system processes products and information. By placing the equipment and people closely together and in a design conducive to flow, many benefits are realized. A cell configuration is never a grouping of the same type of machine performing the same task, because this will only continue to promote batching. One of the benefits of a cell is achieving continuous flow, which reduces inventory and thus lead time. Cells also dramatically reduce transportation waste when configuring the complete process, which again contributes to lead time reduction and supports productivity gains. There is also a quality benefit because with lower inventories the problem is realized and corrected almost immediately, especially when one-piece flow is achieved, and this should be the goal. Quality is also improved when the cell matures to one person running more than one operational step. This single person is now both the internal supplier and the internal customer. With this natural understanding, the operator ensures everything is delivered to the next process step with the known requirements and specifications. If not, it instantaneously causes this single person more work.

Other benefits of organizing the shop floor and office flows into manufacturing cells include process ownership by the shop floor teams, focused process improvements, waste and problems surfacing quickly, and identity for the process teams. Key business metrics like quality, productivity, and lead time improve almost instantaneously in a cell configuration. The mistake organizations make is accepting these initial benefits and not continuing toward continuous flow and problem identification.

Process ownership by the self-directed teams is probably the most significant benefit. Without process ownership it is difficult to get buy-in from 100% of your workforce. The self-directed teams need to get to the point where they feel responsible in meeting the daily customer requirements. When organized as a cell, their scope is much smaller and less complex. Every individual within the team can see both the beginning and the end of the process from one place. They quickly come to realize how all parts of the process are affected by what they do. If their cell provides product for another cell, they realize their finished product has to be produced without defects and on time—the internal customer is now obvious. They also realize they have internal suppliers who provide them with several needs: a product from another cell, raw materials, technical support, and process information.

Manufacturing cells also reduce the complexity of the complete value streams because they take on an appearance of segmented businesses. The flow is continuous, as I mentioned, but because each cell is measuring its scope of the process, improvements and problems surface quickly. I remember when a plant manager who had a very large traditional layout took me on a plant tour and stopped at a board where the unit production output by day was charted. The target was to produce 15 million units per day, and the latest numbers showed more than 16.8 million. He was very proud of this. My question was, what part of the process was responsible for this? He looked at me puzzled, but then realized he did not know because of the complexity of the process. So an opportunity was lost. With a manufacturing cell layout, complexity is mitigated and improvements can easily be realized and shared.

Below is an example of immediate results obtained when the equipment layout was changed into a cell configuration.

PROBLEM

Time to complete a batch was too long to meet the newly proposed daily production goal.

CURRENT SITUATION

The Process Team decided to track 50 batches to determine the average time to complete a batch with the current layout. The 50 batches were sufficient to realize all process variation (Table 2.1).

ROOT CAUSE ANALYSIS

Although more extensive analysis was completed, the critical cause was determined to be the current layout of the machines.

SOLUTION

Lay out machines in a U shape.

RESULTS

A 22% reduction in batch time was realized, and a 41% increase in units processed per hour was realized (Table 2.2).

TABLE 2.1

Average Batch Time, Before

Average Batch Time	Average Units Processed (Hour)
1:44:25	24,775

TABLE 2.2

Average Batch Time, After

Average Batch Time	Average Units Processed (Hour)
1:25:37	34,900

Some suitable results were achieved simply by moving the equipment—nothing else! So you can see it is very easy for management to stop the improvement investment and move on to something else. However, moving the equipment into an effective cell structure is only the first step toward significant fast-paced business results. Further and more significant results in quality, cost, and delivery from the improvements brought by self-directed teams will be sacrificed if management stops here.

MANUFACTURING AND OFFICE CELL IMPLEMENTATION STEPS

The forming of the Management Steering Team is the first step toward a fast-paced organization, but immediately after this the team must go to work on designing and implementing the cell layout. All value streams must go into this configuration to take advantage of the many benefits already discussed.

The major steps are as follows:

The Management Steering Team determines how the new layout (don't forget the office) will contribute to fast lead times, quality, cost, and productivity.

Provide the historical data (current situation) for the metrics, which will determine success, and design the visual management of these metrics.

Identify all of the value streams.

Complete the product and service flow matrix (see Table 2.3) of these value streams. Label all process steps in the columns and the various products in the rows. For each product, go across the row and place

TABLE 2.3

Manufacturing Cell Product Matrix

	Hot Stamp	Mill 1	Mill 2	Press	Oven	Trim	Final
Product 1	X	X	X	X	X	X	X
Product 2	X	X	X	X	X		X
Product 3			X	X	X	X	X
Product 4			X	X	X	X	X
Product 5		X	X	X	X	X	X

Source: Robert Baird.

an X where this product requires a process step. When the matrix is completed, it will become obvious which products can be produced in which cell. For example, products 1 and 2 could be processed in cell 1; products 3, 4, and 5 could be processed in cell 2; product 5 could actually be processed in cell 1 also.

The Management Steering Team determines the U-shape configuration along with the people working in the process and any expertise needed like engineering, facilities, quality, and safety.

Machine inputs like air supply and power are designed for the configuration by facilities and engineering.

The required workstation design is determined by the people who will work within the new configuration. This should now require fewer worktables and cabinets but more quality control stations. It is also advisable to bring in an ergonomics expert to aid in the design.

Most of the stakeholders, because of their participation in the cell design, will already know the reasons and benefits for this change; but some other groups like sales and customer service might not. It is now time for the Management Steering Team to provide this communication to these stakeholders.

The Management Steering Team along with engineering, facilities, and operators put together a detailed project time line for moving the equipment into the new layout. This will include WIP which must be built to provide a buffer for the time that equipment will be down, disconnect and connect procedures, any required cranes, order of machine movement, required workbenches and cabinets, visual management with three-sided boards, and start-up calibration. Considerations should also be given to designing a system where power, air, vacuum, etc. can be easily dropped down to start up a machine from any practical location within the building.

Determine which days the move into the new configuration will occur. (One organization I worked with completed the move in one weekend.)

Remember, the new layout is the first step toward achieving world-class results. Allow the people to work within the new configuration, but then start to implement these next steps:

Start the self-directed teams if not started already.
Put 5S into place.
Start autonomous maintenance.

Develop organizational structure with team facilitators, TWI (Training Within Industry), and Lean Sigma skills.

Implement a career path for the operators.

Implement the supermarket pull system with planning at the Pacemaker.

STREETS AND AVENUES LAYOUT

Having an efficient process layout is a fundamental requirement in achieving world-class lead times.

The streets and avenues layout (see Figures 2.1 and 2.2) is practical for both material movement and the daily tiered Gemba walk. The material (or forms if a service process) storage warehouse can be located at either end of the building or even at the sides of the building. The amount of space taken is minimal, and if the process is manufacturing, the equipment usually fits between the building's load-bearing beams. The streets are where the people move around, and the avenues are where the cells are located.

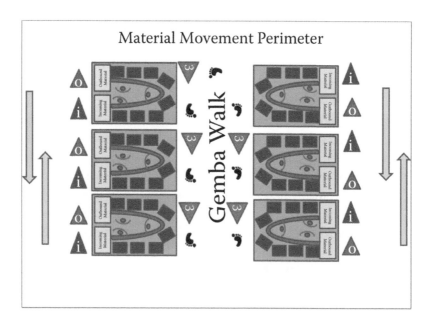

FIGURE 2.1
Streets and avenues. (From Robert Baird.)

Cell Three Sided Visual Management Board

 Incoming Material Storage

 Outgoing Material Storage

FIGURE 2.2
Streets and avenues legend. (From Robert Baird.)

With the material movement perimeter, material movements (raw and finished materials) have reduced transportation time. The arrows show the travel areas and can include the top and bottom of the layout. This type of layout minimizes people transport and material transport conflict, which prevents any possible safety issues. Material transport can actually be set up with a robotic system where the containers are on carts that can be programmed to go to a specific cell and stop. After materials are removed or the container is filled with finished products, the operator pushes a button and the robot goes where it was programmed to go by the warehouse. If the robot senses a person or something in its path, it will stop. This type of system removes considerable waste in the area of transportation by people.

You can see designated areas for storage of product in this layout. These can also be supermarkets for a pull system. Within the cell itself is continuous flow, so no material containers are required. Each cell can be a complete value stream or part of a value stream. You can also see the tiered management Gemba walk is completed in a U shape in the main street. The upside-down triangles represent each cell's three-sided visual management board. This board swivels to make it easy to review each side with the team leader. All of the visual management is manual, updated by the self-directed teams. It is as close to real time as possible, unlike a system-wide LED display system designed by, most likely, engineering. With this Gemba walk flow path and with the correct visual management, the Gemba walk can be accomplished within 15 minutes.

Depending on your value stream you could have more streets and avenues replicating this layout. You could also configure it somewhat differently, but for the most part you must maintain the streets and avenues for efficiency.

SUPERMARKET PULL SYSTEM

Not all production systems benefit from a supermarket pull system, but most do. Why? Because most systems have an operation (process step) that is significantly faster than others or has extended setup time. In this case, there must be a method to link this operation to their downstream process step before overutilization of the downstream operation occurs. For example, if process step 1 produces 100 units per hour and process step 2 produces 50 units per hour, then at some point process step 2 is overutilized—it cannot keep up with process step 1. Then WIP starts to build in front of process step 2. With WIP building, lead time starts to build also, impacting the customer in a negative way. Little's Law rearranged for determining lead time is as follows:

$$\text{Lead Time} = \text{WIP} / \text{Units Produced}$$

This is simple but powerful and can be used to determine the work order flow within your production system by the planning department. Here is a use of Little's Law: For example, if Planning is entering 10,000 units per day into the system (backlog) and you are effectively producing 8,000 units per day into Finished Goods, then WIP is going to build somewhere and so will lead time, again impacting your customers. This effect also directly impacts costs: With longer lead times you need more space for inventory, loss of revenue causes cost per unit to go up, and labor hours are a premium because of overtime required to meet on-time delivery dates for your customers.

Little's Law is a powerful and simple formula to direct you toward managing the WIP levels. However, as we know, our production systems are very complex, so it cannot be used to set the required WIP levels. A study of variation can provide a starting point of current WIP levels required. Once we load to these levels, we will then be able to have a more predictable lead time on a daily basis, which helps customer service provide more accurate completion dates for customers. A manufacturing resource planning (MRP) system is not the answer. Process output variation prevents an accurate MRP system, so it would have to be updated almost every minute of the day.

There is nothing better than the combination of visual management and standard work for managers applied with the Gemba walk to manage the designed WIP levels. Without visual management the supermarket system is difficult to sustain. Each supermarket contains inventory containers loaded with each product family. The containers have visual maximum and minimum levels. After your variation study of the daily outputs of each of your process steps (should be completed over a two-month period or longer) or if you have representative historical data, you can determine the required maximum levels. With the maximum levels determined, you can then determine the maximum lead time your system will have. If the lead time calculated is unacceptable, then you now know which process step variation needs to be reduced to lower the WIP level. The Gemba walk is then used to monitor the WIP levels. Are certain inventory containers at process steps always close to running out? Are certain inventory containers at process steps always with high WIP levels? These questions are easily answered with visual management. If they are always close to running out, then the maximum-level indicator needs to increase; and if they are always showing high WIP levels, then try to lower the maximum-level indicator. With the supermarket system you can also record the number of Kanbans required within the system. The objective, on a continuous basis, is to get the WIP as low as possible without affecting the output of any process step. The system is also very beneficial in determining where process variation is too high. An inventory container that is constantly running out or close to it can be indicative of high process variation. We now know where to concentrate our improvement resources. The visual management along with the Gemba walk is very effective in managing the WIP and therefore the lead time and therefore customer delight.

Functionality of Supermarket System

The downstream process step withdraws what is needed and when. The withdrawal cards are supplied by Planning at the beginning of a shift, and either the cell self-directed team goes to the supermarket and retrieves these work orders or there is a Spider doing this for several cells.

When there is a withdrawal from the supermarket, a Kanban is sent back to the upstream step (the step producing to the supermarket) for a signal to start producing to replenish the inventory container the withdrawal came from (see Figure 2.3).

The Planning Department plans at a single point, usually at the Pacemaker step in the process. There is no need to schedule the upstream process as the withdrawal and Kanban system will accurately control this process step.

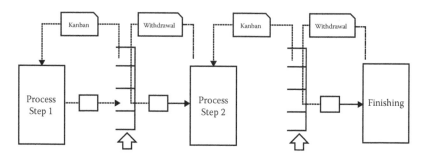

FIGURE 2.3
Supermarket system. (From Robert Baird.)

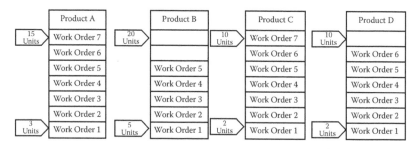

FIGURE 2.4
Supermarket example. (From http://www.vision-lean.com/leantek-applications/lean-manufacturing-supermarkets/.)

Because the upstream process is faster, there will probably be some downtime because of no Kanban cards to signal production. This makes the traditional production people very nervous, and it will take some time to get over this anxiety. This is another benefit of the tiered Gemba walk; managers can monitor the inventory container levels and use their teaching and promoting techniques of STP (support, teach, and promote) to prevent any overproduction to the supermarkets.

You might only require one supermarket (between two steps) in your production process. If it is only one, the downstream processes should be in continuous flow (one-piece flow). This requires the downstream processes to be in manufacturing cells.

For an example of a supermarket, see Figure 2.4.

Supermarket Implementation Steps

The supermarket pull system is used when continuous flow cannot be achieved. Design and implementation is another responsibility of the Management Steering Team. Here are some suggested steps to put into place:

Start with the Management Steering Team designing communication to all of the stakeholders, including planning, self-directed teams, customer service, sales, supply chain, possible suppliers, team facilitators, and warehouse. The main message of the communication is to provide world-class lead times to satisfy customers and that each and every position is responsible to make it work—that's it!

Generate or update your current value stream map.

Record customer demand of each product family.

Calculate the required Takt time.

Measure your current levels of WIP at each process step.

Measure the process step setup times. Excessive setup times must be resolved before using supermarkets, or the replenishment times will be too long, requiring larger amounts of WIP, which of course increases the lead time.

Accurately calculate the cycle time of the process steps and the overall lead time that includes the backlog but not the work order generating process. Please use historical data if you have it.

Measure the variation (standard deviation) and use a histogram for the output of each process step.

Determine the daily demand from historical data for each of the product families and then calculate it for each shift; this is the WIP level you will be working with.

Measure the variation (standard deviation) with a histogram for the demand of each product family.

Graph seconds/unit of each process step (sum all machines within this process step) and the Takt time determined from the customer demand. This will give a visual of where some process steps are much faster than others and may require a supermarket. It will also show you where you are either able to meet Takt or not able to meet it. Process improvements or additional resources may be required if your process seconds/unit is higher than your Takt.

Generate a Pipeline Map as follows (see Figure 2.5):

Measure the actual daily output of each process step.

Measure the instantaneous speed of each machine type by observing the output over five minutes without interruption.

The actual daily output is plotted on a line graph.

Then units/hr (determined above) × available production hours in a day is plotted on the same graph (top line).

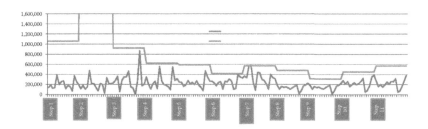

FIGURE 2.5
Process map. (From Robert Baird.)

This will provide another visual of where a supermarket might be needed and the added benefit of where a process might be overutilized, indicating a bottleneck. In Figure 2.5 you might need a supermarket between steps 8 and 9.

For the process steps you selected for a supermarket, you can now calculate the inventory container quantity for each product family by the following formula:

$$(Do + SD \times Uct) / C$$

where:

Do = average shift output (units) of downstream (or customer process) step

SD = standard deviation of shift output (units) of downstream step

Uct = cycle time of upstream (or supplier process) step. (Replenishment time is once the Kanban signal has been received.) This measure is in days, so if the upstream process has a cycle time in hours, it is hours/24.

C = Number of containers for this product

For products that are not being produced each day, the upstream process (supplier) can produce to its inventory container when they have some downtime from the regular products. Planning will provide the date to be completed.

Kanban and withdrawal card boards are set up to make the process even more visual.

The withdrawal cards are placed according to estimated time to pull from the supermarket. These have some of the required work order information. This is Planning's daily responsibility.

Once the upstream process (supplier) has seen or been signaled with a withdrawal of a certain product type, they can start producing according to work order flow requirements. If it is a made-to-stock operation, they do not have to pay attention to work order flow requirements.

It is now the responsibility of the manager via the Gemba walk to manage this supermarket pull system; some adjustments will be needed to container WIP levels and support for people getting used to how to work with the new process.

Metrics like daily lead time, WIP, and daily output to finished goods must be graphed and be part of the visual management reviewed by the manager's Gemba walk. Remember, increasing WIP will increase the lead time, and we must be careful to not lower this to a level where we might reduce the daily output. This can happen if we sometimes are running out of WIP at certain operations, especially at a bottleneck.

LEAD TIME REDUCTION—CUSTOMERS DEMAND IT

Without acceptable quality an organization will suffer in the market, and the associated costs of poor quality can be extremely high. Organizations have spent the past 70 years improving the quality of products and services to where it is not a market differentiator, but it is expected. Today, we as consumers are now enjoying these higher levels of quality. Even when there are quality issues, like when our car breaks down, our credit card is scammed, or a flight is delayed because of mechanical reasons, the provider of these products and services will adequately compensate. But let's think about when we experience a quality issue: What it has really taken from us is time.

The new measure that determines who organizations select as a supplier is lead time. If an organization has a proven track record of fast and on-time delivery they are usually the leaders within the market. Lead time is the one metric that is indicative of sound and proven processes. It tells us that quality is high and costs are low. In the previous chapter I mentioned the concept of a single-focus strategy; lead time must be the single-focus strategy. Quality and costs are inputs into lead time; you cannot have world-class lead times with poor quality and high costs.

Here are seven of the main steps to take to improve your lead time to world-class levels:

1. Measure it. Many organizations are unsure of their exact lead time. How long does it take your organization to process your customer's order from time of receipt (purchase order received) to the customer receiving the product or service? Even if the organization has a lead time metric, it is customarily measured on a monthly basis and it is an average of a variety of processes. This higher level measurement is in effect a waste as this diluted monthly number is unable to lead the organization to where they must focus problem-solving resources. It is also a waste because even though it might be communicated each month, very few serious actions are ever taken. So, what do you measure to be confident your problem-solving resources will have an opportunity to be fast-paced in lead time reduction?

   a. Break down your lead time into cycle time of each process step. Organizations that provide only services can also break down each process step into cycle time. Graph, with a line chart, the cycle time data by process step for each day of the month. The X axis will be the days of month and the Y axis provides each process step cycle time. This will provide a visual of which process has the highest cycle time. For manufacturing organizations, include the transactional processes like customer service and work order generation. Remember, the customer feels the complete process lead time.

   b. Once you have the above breakdown, the highest cycle time process steps are again broken down by the various types of products or services. This will determine which product or service you will need to start first with process improvement.

   c. Measure daily the output and input of your value streams. If you are, for example, inserting 10 units per day and the output is 8 units per day, the system is going to build WIP somewhere—increasing the overall lead time. This will help determine the true capacity of your process and how much work to start each day.

   d. Graph the daily unit output of each process step. The X axis will be at least 14 days in a row for each process step and the Y axis is the number of daily units produced. The second line on the chart is daily instantaneous capacity of each step. This is determined by observing and counting each process step's output for a short period and then calculating it into a daily output. This will determine two important characteristics of the process: which process step or steps are overutilized, indicating the bottleneck, and which process step or steps are underutilized, indicating an opportunity to move resources to support process steps with higher utilization.

e. Measure the WIP within the process, again daily. Along with the daily output measured at each process step, you can use Little's Law to determine a constant WIP level to match the targeted lead time. Measuring, with the aid of visual management, the WIP each day can inform the value stream workers when WIP is too high or is getting dangerously low. This will enable them to make corrective decisions at a much faster pace. Remember, WIP is not only related to manufacturing, but WIP can be orders waiting to be completed within a computer system, people waiting to see a doctor, cars waiting in a drive-through, and so forth.

f. Determine what an acceptable standard deviation is and measure it daily. Most of us have heard about the problems associated with process variation, but how many organizations actually measure it? With the measurements stated above, the standard deviation can be determined. Are the output variation from each process step and the output from the complete process acceptable? Does the process meet both the internal and external customer requirements on a daily basis? Is the process sometimes overproducing (increasing WIP and lead time) and sometimes underproducing (increasing waiting time and lead time)? Determine which process step has the highest variation and improve it.

g. Measure changeover times. Some processes have long changeover times and are impacting the process flexibility and capacity.

2. Design the process with Process Step Cells. Cells in both manufacturing and transactional services provide the opportunity for *one-piece flow*, zero WIP between process steps. I once worked for an organization that had the starting process step on one floor and the second and third process steps on another floor. To improve the lead time, we simply moved everyone working within these process steps closer together to form a cell. With this change alone we improved the process lead time by 50%.

3. Remove the requirement of supervisor approvals. Approval reviews of someone else's work are only an indication of a process with inherent errors and management using the old command and control methods. I am old enough to remember the supervisor approval required when a grocery order exceeded a determined amount. This frustrated the customer and increased process waiting time. But it

was obviously wasteful because the supervisor would come to the register, quickly review of the person's check, and then approve it. What value did they provide? Could the cashier not be trained, through standard work, for this approval skill? That is exactly what they did, and cycle time improved. One other technique, to take the step toward not needing approvals, is to use Poka-Yoke so that human error is eradicated from the process step.

4. Cross-train. From your measurements you will find that some processes produce faster than others and some are bottlenecks. Some process steps are so much faster that people spend most of their time actually waiting. Cross-train people to support slower processes. The need for cross-training becomes obvious when the process is organized in Process Step Cells.

5. Reduce processes with long changeover times. Some process steps will have adequate capacity and low variation but long changeover times that, in effect, reduce capacity. Also, it is impossible to achieve the benefits of one-piece flow with long changeover times. During your value stream analysis, look for these longer changeover times and determine if they are impacting the ability for continuous flow and processing small batches.

6. Remove non-value added steps. If the process can eliminate a process step, mainly through technology, you immediately remove the cycle time of that process step. I do not advocate removing process steps through outsourcing; in most cases this only adds to lead time and the organization then becomes too far removed to effectively improve these cycle times. Quality checks can also be, in effect, a process step with a significant cycle time. Removing these quality checks by building quality in upstream can remove noteworthy cycle time.

7. Balance the work order mix. Many organizations still believe in sequencing the large orders first and then attempting to schedule smaller orders when they can, when they find capacity. This never works! The longer processing times simply absorb most of the capacity. People usually do this because of unacceptable changeover times. If changeover times are reduced, the orders can be sequenced by their associated customer request dates. This improves on-time deliveries.

Other steps can also benefit lead time reduction, but these seven proven steps provide the most value. Process design techniques can also be used, like applying the theory of constraints, installing supermarkets, and applying Six Sigma to support lead time reduction. Learn these valuable techniques and apply them where applicable.

Take on lead time as your single-focus strategy and become the leader within your market.

EVERYDAY EXAMPLES OF USING LITTLE'S LAW

The analogy of automobile traffic and manufacturing production flow has probably been used too many times. I have come to this conclusion because I never hear it used anymore. However, it is a very good analogy especially when you look at the simple formula of Little's Law. The flow of traffic has many elements similar to most of our manufacturing production flows. Traffic can flow through one lane or many lanes, just as production does through one or more lines or cells. Traffic has downtime when there is an accident, and the reaction and repair time to get the lanes back to normal flow can be considerable (I often wonder why). Traffic has special orders entered into the system when there is a presidential convoy, for example. Traffic has a significant amount of variation that manufacturing flows share in human beings and materials. Traffic flows are planned by civil engineers, who use some of the same calculations used in manufacturing, such as capacity, utilization, and some controls.

Many states now use traffic lights on ramps that enter a busy freeway during peak periods of flow. They do this because of what Little's Law tells us. If the capacity of the freeway or utilization goes beyond 90%, then because of variation (unfortunately we cannot predict the driving habits of humans), we will only build WIP and cycle time increases. So the civil engineers try to manage the utilization of the freeway.

The next time you are driving to work, try driving the speed limit and staying within a single lane (do not change lanes). Time your trip with this approach, and then compare it to your normal approach of traveling, somewhat over the speed limit and frequently changing lanes. You will have to take a sampling of about 10 trips, but you will find that "slow and steady" does get you there with less cycle time. Why? Little's Law again—going over the speed limit is exceeding utilization and changing lanes causes variation not only to your journey but to the other people fighting for capacity.

Many who travel know very well the need for air traffic controllers to not overutilize a runway. This is something we take for granted, and the controllers (the planners) know very well to actually underutilize runways or disaster is imminent. The other airport analogy of flow is when the plane is leaving the gate. How many times have you left the gate for the runway only to hear from the pilot, "Looks like we are number 26 in line for takeoff." The control tower gives the clearance for a certain runway and in a certain order. If the pilot does not go to the runway or goes in the wrong order, it will be a much longer wait. This is very similar to manufacturing; the planners provide a schedule, but once it gets to the shop floor, the first-line supervisor rearranges it to get things moving! Also, the planners send work orders to the first operation hoping that the sooner they get it in the queue the faster it will be produced!

Little's Law holds true in many processes of flow, and we can certainly take advantage of this simple formula in manufacturing planning of flows.

In the dice game that became popular from Goldratt's book *The Goal* (1984, 2012), there is a round where the first station is allowed to introduce only the volume that came out of the last station. The resulting measurements from this round are less WIP and subsequently faster cycle time. This is the essence of Little's Law: overutilization of the production system only builds WIP and cycle time.

An age-old saying related to quality is, "Garbage in, garbage out." As we all know, this is an axiom worth paying attention to. I have started a saying related to planning the schedule according Little's Law: "Quantity out equals quantity in." We must pay attention to this. It will definitely guide you to where improvements can be made. If you are unhappy with the daily volumes going into finished goods, either your WIP is too high or your production system capability is too low. Yes, we need to improve the bottlenecks, but we must also control WIP and the variation.

PROCESS METRICS

We have to look at what we measure in terms of how it affects our actions and how the results connect to our internal and external customers. Most companies severely underestimate these two responsibilities. All levels of the organization must be able to understand their indicators of success. All levels must be able to connect their efforts toward meeting their indicators to the single-focus strategy. There must be a line of sight to the overall strategy. This is obvious but not so easy to accomplish. Consider the self-directed teams. We as managers and engineers want to measure global and high-level results of our processes, the output measures. At this level these metrics mean nothing to the shop floor—ironically those who are the direct inputs for these global metrics. The self-directed teams understand and influence input metrics. The self-directed teams will be more effective when they have accountability for process input metrics and not output metrics. Input metrics like their 5S score (discussed in detail in Chapter 5), variation of heat or pressure, WIP levels, shift change time, setup time, training matrix, and others can be used. I used to ask plant managers and engineers what operators have to do better to improve cycle time, but most of them could not provide an answer. This is because managers think in terms of and are more familiar with output measures. Performance metrics must cascade throughout the organization, and what people measure influences their behavior. At the highest level the metrics start as process output measures, but as they are cascaded they transform into input measures influencing these outputs. A simple way to determine the required input measures is to use the Six Sigma SIPOC (suppliers, inputs, process, outputs, customers) tool. As leaders you must understand what needs to be measured by the self-directed teams in order to influence the single-focus strategy, and your people must know how they can contribute.

If the shop floor self-directed team metrics are the right ones, it will be obvious. The following behaviors will be realized:

Drive the cell team behavior of improvement.

If a metric-related task is changed, the metric will change. For example, if you are measuring the cell cycle time and an operator within the cell changes a procedure (installing a die, machine setup, flow change, etc.), it will be seen immediately in the cell cycle time metric.

If the shop floor metric improves, the next level metric should improve (direct relationship is a must). So if you are measuring yield loss at the cell and the self-directed team improves yield loss, the global metric of cost of poor quality should also improve.

The entire cell team, 100%, clearly understands the metric. Each team member understands what they do can directly change the metric.

In all cases the format of the metric (type of graph) must be able to show overall trend and variation.

To establish metrics we must first start with an enterprise single focus. What is the overall metric that, if improved, will require that other metrics improve also? Lead time is one of these metrics; quality and productivity cannot be poor if you want to have good lead time. Cost can also be a single-focus metric, but be careful because cutting direct labor is easy. Traditional managers think of labor as an expense but machines as an investment; it should obviously be the other way around. After the single focus is established, start obtaining the support of various enterprise structures (departments and self-directed teams). These structures should in turn determine more direct measures to achieve the single focus, inputs influencing the outputs.

To confirm the progress and continue momentum, Standard Work for Managers checklists are developed for checking the results of these measures during the daily Gemba walk. The metric development, STP of the single focus, is the responsibility of the Management Steering Team. Each metric is assigned to a project owner and follows a proven problem-solving methodology. Self-directed teams are project owners and have accountability for their contribution. Of course these global and high-level metrics have some value, but we need to understand the line of sight for every project within the plant. Please reference Team Process Improvement Metrics section in Chapter 3.

Processes must be measured, but the more critical element is communication of these output measurements. Two main internal customers require this communication: the shop floor and the sales group. As we gain results, no matter how small, we must communicate this momentum to these two internal customers. Momentum of gains is powerful, but without clear communication the momentum stops, because in the early stages of improvement programs the gains are not obvious. These small gains need to be communicated to get all internal customers motivated and supporting the improvement projects. Celebrations of early gains are also very important in maintaining momentum. People want to be winners, but they will quickly lose interest if there is no connection with the score and their work.

Review of the metrics, especially the process metrics, must be consistent and frequent. Metrics need to be reviewed daily by the self-directed teams and through the tiered Gemba walk. If it does not make sense to review them daily, then you are measuring the wrong stuff. The process is dynamic, and the self-directed teams are seeing the effect of these dynamics.

Managers and engineers usually want to have the metrics automatically retrieved from various databases in order to remove redundancy, inaccuracy, and wasted time of the operators updating the data—but this is the wrong thinking. Operators will not take ownership of data given to them. When I see automatic data for the process metrics, I immediately question if they are the right data; can input metrics be automatically retrieved? Measuring materials in a Kanban, the number of times a machine stops for material shortage or breaks, the root cause of a specific defect, the number of TPM events completed by the maintenance technician, cycle time of a key machine operation task, shift change time, setup time, overall equipment effectiveness (OEE)—are all examples of process metrics that must be manually gathered by the self-directed teams. An enterprise resource planning (ERP) or other database system cannot get these numbers. If you purchase a systemwide data collection system, have you planned for the additional resources to maintain it? The manual updating of metrics requires shop floor language and is related to quality, cost, and time; these are the fundamental measurements for manufacturing and other processes, including service.

Celebration of success certainly helps maintain the momentum of process improvement. I used a very effective celebration program to obtain everyone's support in striving for the results. When the self-directed team posted the graphs or tables, I would have them clearly identify three different levels of achievement (you can use more, but three is the minimum). We would then predefine what we were going to do when we reached these levels. For example, Level 1 could be a pizza party, Level 2 a dinner at a nice restaurant, and Level 3 a monogrammed wristwatch. We would also identify where World Class was on the charts and how we were going to celebrate once we reached it. Reaching World Class was of course a very special celebration with events that involved spouses and families.

QUALITY SYSTEM

In traditional manufacturing environments, quality inspections of produced product usually occurs at receipt of goods and through sampling after a product has gone through several operations. In Lean manufacturing environments, inspections are accommodated at any point in a product's life as it is being produced. But the inspection is of known conditions to produce a defect-free product. The continuous improvement process is to find the key points and key process inputs so that defect-creating conditions can be controlled, minimizing variation. TWI or Training Within Industry consists of four programs, TWI-JI, TWI-JM, TWI-JR, and TWI-JS, providing front-line employees with the skills to provide improvements and change. The programs uncover the key operational points which in the past were only gained by experience. The key points are integrated in the standardized work, with everyone in the process trained on them, and the key process inputs are controlled with tools like SPC or Poka-Yoke. This is Lean Quality Control.

To find these key points and key process inputs I have developed a process called Zero Quality Discovery. This process is outlined and explained in Chapter 5, "Lean Sigma Tools."

Zero Defect Process: What Stops Us?

In a past LinkedIn Six Sigma group, people were providing arguments for not going beyond a level of quality expected by the customer. The main argument was, why increase your cost for a level of quality your customer does not require? Think about this argument: Is it not revealing an organizational culture? More on this in a minute.

I have asked different organizations to tell me what an acceptable level of yield loss was. It always surprised me that they would answer somewhere close to what they were currently at (unless they currently had a single-focus strategy of improving it). I even had managers of an organization reply that 8% yield loss was an acceptable level! Yes, the 8% was their current yield loss.

What are the implications of accepting a yield loss level of even 1% (or 10,000 ppm). First, a process producing any level of defects presents the possibility of a Type II error: in statistical terms, the failure to reject a false null hypothesis; in layman's terms, a failure for the process to reject a defect that therefore is probably going to your customer. How many customer complaints do you have related to Type II errors? What is the cost of these complaints? The loss of a customer is extremely expensive; some people put this in perspective by stating that it costs five times as much to attract a new customer as it does to keep an existing one. If your organization has a yield loss greater than zero, then most likely you incur inspection costs. Most likely you are starting more material than the actual order requires to fulfill the requested customer order volume—a cost you incur. Most likely you install more capacity and manpower than a zero defect organization. With a yield loss greater than zero, you have more people in the quality department than a zero defect organization has.

What stops us from reaching a zero defect process? One argument I already presented is used by many people: Why increase your costs for a higher level of quality than expected by your customer? However, this is not really true when summing all of the costs I mentioned above. But even more than this, it is an organizational culture. A culture of acceptance, a paradigm of expecting defects, it is the norm, and we see it every day. Take a test of your culture (you might have already done this) and get the decision makers together to propose the question, "Can we achieve a zero defect process?" What is the response, and what are the reasons why this is impossible? I worked for an organization that was in the fourth year of a Lean journey. We accomplished some world-class results in productivity, cost per unit, and delivery. Even in quality we achieved levels much better than our sister plants, but not a zero defect process. So our Management Steering Team decided to embark on reaching this target. This was a high-speed process producing an average of 3,000 units/hour. The first step was to assign a diverse work team to work full-time on reaching the target. Because we were organized in self-directed teams, 80% of the Zero Defect Team consisted of team members whose normal job was to work directly in the value stream. Within six months the team developed a repeatable methodology, labeled Zero Quality Discovery, for achieving zero defects—yes, we actually achieved zero yield loss.

When we started this endeavor, the organization already had a continuous improvement culture but not specifically a zero defect culture. So when the question was asked to the Management Steering Team, "Can we achieve a zero defect process?" there was some doubt. This doubt challenged our existing culture, bringing us to another level. We also did not need a "burning platform" scenario to motivate us to take on this daunting task. We had the benefit of already having a well-developed continuous improvement culture, which enabled the movement of our paradigm.

Develop your continuous improvement culture and take on daunting targets to move your organization into a leadership position.

LEAN QUALITY CONTROL

A definition of *quality*: meeting customer requirements (stated or implied) through the totality of a product or service by conforming to a specified standard at a given time over a period of time at a price the customer can afford and is willing to pay for.

Lean quality control is all about controlling input variation. Controlling quality at the output of the process does not lead to quality assurance; some defects will always escape. Many Lean Sigma tools can be used to accomplish Lean quality control: SPC, Jidoka, process supplier and customer system with self-directed teams, supplier partnerships, Scrap Box Reviews during the Gemba walk, Zero Quality Discovery, and Poka-Yoke. I will provide a discussion for each one of these valuable tools.

The basic requirements of Lean quality control are as follows:

Design quality in
Process input control

Statistical Process Control

SPC has been around since the 1920s, but in most cases has been implemented so poorly that it is unable to detect the special causes of variation that can negatively impact the process. As a result, management will usually make the decision to remove it or not provide any support. If you want to remove an expensive inspection department, you will need to correctly implement SPC. Remember, the key tools within SPC are control charts, design of experiment, and continuous improvement. They are for identifying, reducing, and eliminating process variation. SPC is not only the application of control charts. The value of using these tools is in having a stable and therefore predictable process. The concept behind the correct application of control charts is to prevent an out-of-specification condition by identifying the out-of-control condition and then correcting it. This customer protection method is something your process cannot be without. This is the prevention part, but the ultimate objective is to reduce or eliminate the sources of variation—and this is where design of experiment and continuous improvement are required. SPC must be looked at in this way, as a complete continuous improvement program aimed at improving process capability. I have provided further explanation and implementation steps for SPC in Chapter 5, "Lean Sigma Tools."

Jidoka

Jidoka is autonomous quality, which means everyone has the responsibility to deliver quality to their customer. The customer is mainly the next step in the process (internal customer), but the external customer is always implied. You have heard of *customer driven*—what does this mean to the workers adding value to the product or service every day? Do they have the required customer knowledge and recent concerns to make everyday decisions on quality for the external customer? These decisions can of course be customer driven with variable data, something that has clear, measurable specifications from the customer. But they are not customer driven decisions when it comes to attribute data, like cosmetic issues. For example, a certain customer received scratches on their product three orders ago. First, does the worker making the decision know the details of this complaint? Probably not. And if they heard, do they remember? Next, the worker is now at the point when they see a scratch for this same customer, which is a Type II error (it was not rejected when it should have been). This error is not the fault of the worker—it is the fault of the system, which is not customer driven.

Process Supplier and Customer

One program I had tremendous success with is part of the dynamic problem solving process (discussed in Chapter 3, Dynamic Process Improvement section) and promotes being customer driven: the process supplier, process customer program. For each key step of the process we would clearly identify the supplier of quality and the process customer. For example, the process adding an attribute to the product and passing it to the next process step is the supplier of quality; the customer is the next step in the process, which might have to apply heat and pressure to this attribute. Each customer in the process would identify the specifications required by the previous process step. Most of these requirements are already known from quality and engineering, but through regular communication (weekly meetings) with the process supplier and customer, you would be surprised what other requirements become inputs for quality. It could be that a specific area of the product must be kept clean, or no overlap of material applied because of machine jams in the next process, or the type of product transfer container needed to prevent scratches. If you have a self-directed team organization, this program becomes almost natural.

During the process supplier and customer weekly meeting, which lasts no longer than 15 minutes, the two teams get together and review the critical-to-quality issues realized in the past week. These issues are manually tracked and visually managed by the customer in the process. In some cases this meeting is held more frequently because an issue is becoming more frequent, so the customer might call a meeting to understand how it can be resolved by the process supplier. And the customer in some cases will have some potential solutions to offer to the supplier. Any solutions addressing root causes realized are the responsibility of the process supplier to effectively address and become standard work. With self-directed teams these meetings do not require the authorization of a supervisor or team facilitator because the team is empowered. Once this process supplier and customer relationship has been working for about six months, you can also start introducing the direct handling of external customer complaints. If the customer complaint comes from this part of the process, then the self-directed teams should be responsible to resolve it and provide the report for the customer. This program provides a fast-paced resolution for quality issues. Root causes are realized because they are fresh; we are not using historical data here. This is natural when you have continuous flow within a cell; an individual within a cell can be both the process supplier and the customer if they are skilled to simultaneously run two machine types or process steps. This is the best scenario because the worker, in real time, discovers what must be done to prevent poor quality and increased cycle time. The key in this situation is again to identify the key points so that it becomes part of the TWI-JI breakdown and then it becomes standard work for everyone. The benefits are less process variation and increased quality and productivity.

Purchasing Department and Supplier Quality

> End the practice of awarding business on the basis of a price tag. Instead, minimize total cost. Move towards a single supplier for any one item, on a long-term relationship of loyalty and trust.
>
> **—W. Edwards Deming**

McDonald's restaurants achieve success through a focus on mutually beneficial collaboration in the supply chain. McDonald's is not concerned with saving a few pennies by continuously switching suppliers, or squeezing their suppliers for the "privilege" of selling to McDonald's. They are focused on developing a long-term relationship with their suppliers to develop mutually beneficial collaboration opportunities. They have proven the value of this model; they have been extremely successful and have managed to maintain low prices for their menu while developing relationships with their suppliers. There are not too many companies who can proclaim these achievements. This model is counterintuitive to most managers and especially purchasing departments, who have alternatively developed a model for low-cost materials. This model creates playing games, resulting in undermining quality and delivery.

Is your purchasing department involved in your continuous improvement program? If the answer is no, this must change. Most organizations would answer no to this question because this department is either in a different location or part of a corporate vertical not directly reporting into operations. When this is the case, they are also usually the only ones talking to the suppliers on a consistent basis. This is concerning for two reasons: They are not close enough to the value streams, and their focus is on getting the best price first and foremost, delivery next, and quality is down the list. The other common issue is that they continue to renew contracts stating material specifications that have been there for years. I have seen contracts where the specifications had not changed for more than 14 years and even engineering did not know where the specification limits came from! This is not indicative of a continuous improvement program. When the purchasing department is questioned about these longer term specifications, they will argue that if specifications are changed, the supplier will want more money! Operations is very familiar with how much process variation is introduced by raw materials, so we should, at minimum, be asking for improved Cpk values. The Cpk values of each raw material will provide operations insight as to how the materials will run or adjustments which might have to be made to their process. Another problem with the mostly singular relationship between the purchasing department and suppliers is the purchasing department will usually defend the supplier when there is a problem and the usual resolution is providing another batch. There is typically no complete problem-solving methodology followed by the supplier, so operations will usually see the problem again.

How is this resolved? Operations should take the lead and get involved in the new supplier contract. I am sure purchasing would welcome local operations to review the specifications. This review must be data driven. Engineering must do some capability studies or even design of experiment (DOE) to determine how much variation is acceptable. Operations should also provide the manufacturing cost information impacted from variation. This can be completed by simply calculating and charting the Taguchi loss function. These are the responsibilities of manufacturing operations.

Finally, partnering with key suppliers provides significant value and facilitates a continuous improvement program. I have found the best method to start this partnership is to provide not only the objectives and quality improvement targets for your company but also the benefits for the supplier as part of the partnership charter. You are creating a mutually collaborative relationship. For example, for the supplier you might state a longer contract, share in the quality gains, and enable them to share some of the material improvements gained from this partnership with your competitors. The partnership we developed requested that each company put together a team; our team consisted of quality, engineering, operator, and purchasing. We also shared meeting locations, so one month the meeting was at their location and the next at our location. We always started off each agenda with a tour of the manufacturing process. The tour provided many benefits and an even better understanding of the material limitations both the supplier and our team had not realized. I'll always remember the comment from one of the supplier's team members, "I never realized our product was subject to this level of pressure." This realization alone improved the product with a change in material formulation. We also established process control charts for the key material characteristics within the supplier's process. In the beginning we had to train them on the use of these control charts, which was appreciated by the supplier. With every order, they would send a copy of the process control charts. This again realized early benefits. For example, they once sent a control chart that was not out of specification but was out of control, with seven consecutive data points below the process average. So we ran this material batch on one of our production machines and realized a breakthrough in adhesion. This became a new specification that our organization benefited from globally. This would have never been discovered without this type of program and partnership.

Scrap Box Review

Another program I use is what I call Scrap Box Review. It is a Gemba review of what the operators have declared as a defect and discarded to a box or bin. Managers over the years have tried many different methods to teach the workers, who make day-to-day decisions, what is acceptable and what is unacceptable for product or service quality. I have seen defect catalogs, LED displays of what is unacceptable, examples hanging on walls, and of course training. All of these provide positive effects, but the problem is they are too static. The customer's implied quality can change with time and from various people within their organization, especially when you are introducing a new product. The managers are the first to know about any changes in what is acceptable by the customers. Is it effectively passed on to the workers? The answer is no. The knowledge passed on might be OK for the day the training happens or the new displays go up, but within a short period it is mostly forgotten. How many repeat complaints do you have? With the Scrap Box Review process, workers are updated and reminded on a daily basis. Here is how it works.

> During the manager's daily tiered Gemba walk, the supervisor's or team facilitator's checklist has a standard check of observing the scrap the workers are rejecting. They also review the products they are not rejecting.
> From the observations, they determine if any Type I (rejecting a good product) or Type II (keeping a reject) errors are being committed. If yes, the supervisor teaches the worker on nonconformances.
> The other check at this time is to determine if sampling is according to standard work.
> At the same time, a quick Kaizen occurs if rejects are above normal levels.

This is a great method to support, teach, and promote (STP techniques) product or even service quality.

Zero Quality Discovery

Product defects hurt the company's reputation with its customers and also waste valuable resources in scrap. Companies that pursue low-inventory production no longer have a large buffer to absorb quality defects. To keep production moving smoothly, it is especially important to prevent defects.

Zero Quality Discovery is a Lean Sigma tool I developed to identify key points and key inputs required to produce zero defects. The improvement process is completed during live production, and this is one of the benefits—no lost production during the event. The benefit of live production is the ability to immediately discover the root causes of defects. Once this is discovered, the operation becomes aware of key points and key input management to prevent these root causes from occurring. I provide the methodology in Chapter 5, "Lean Sigma Tools."

Poka-Yoke

Because people can make mistakes, even in inspection, mistake proofing often relies on sensing mechanisms called Poka-Yoke, which check conditions automatically and signal when problems occur. Poka-Yoke devices include electronic sensors such as limit switches and photoelectric eyes, as well as passive devices such as positioning pins that prevent backward insertion of a work piece. Poka-Yoke devices may use counters to make sure an operation is repeated the correct number of times.

The key to effective mistake proofing is determining when and where defect-causing conditions arise and then figuring out how to detect or prevent these conditions every time. Shop floor people will have important key points (TWI) and ideas to share for developing and implementing Poka-yoke systems that check every item and give immediate feedback about the problem.

The best Poka-Yoke device is a mechanism that does not require intervention from the operator, like a software feedback loop automatically adjusting the depth of a mill before it goes out of specification. The next best device stops the machine when a known condition arises. The next best device is an audible or visual alarm signaling the operator to correct something. All of these Poka-Yoke mechanisms should be configured so that the defect is actually prevented.

So what causes defects?

- Cultural factors
  - Wrong procedures or standards
  - People expecting and accepting defects
- Variance
  - Material variation (thickness, finish, etc.)
  - Worn machine parts
  - Input drifting
- Complexity
  - Too many steps
  - Many input variables
  - Unique knowledge
- Mistakes
  - Humans—the most frequent cause of defects

What are the Lean quality tools to prevent or eliminate defects?

- Cultural
  - Management commitment and support
  - Empowered teams accepting process responsibility
- Variance
  - SPC
  - Six Sigma methods (having access to and using process data)
- Complexity
  - Process mapping
  - Technology—sensors
  - Source inspection and TWI for identification of key input variables
- Mistakes
  - Mistake proofing, Poka-Yoke

Human Error

To err is human! Have you ever driven to work and not remembered certain events? or driven from work to home when you meant to stop at a store? This happens to workers also; workers finish the shift and don't remember everything they have done. After building green widgets all morning, the worker puts green parts on the red widgets in the afternoon.

What are the common corrective actions? Unfortunately too many managers fall into easy decisions of worker reprimand and retraining.

The warning to "be more careful" or "pay attention" is not effective for humans, especially in repetitive environments. A new Lean quality culture must be developed:

- Stop blaming the human. Correct the system!
- Make wrong actions more difficult or impossible.
- Source inspection versus judgment inspection (see Figure 2.6).
- Make it easier to detect the errors that occur.
- Make incorrect actions correct.

One Management Steering Team I was on decided the only way to actually achieve a quality level of zero defects was to form a team of our best team members and to dedicate 100% of their work time to reaching this daunting goal. I was assigned as the team facilitator, and my first responsibility was to select the team members. I decided to select a person from each of the process steps who showed the passion for reaching the assigned

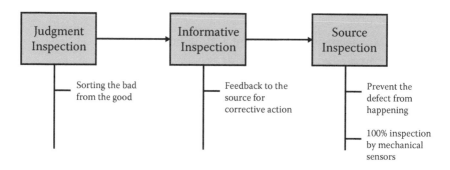

FIGURE 2.6
Inspection evolution. (From Robert Baird.)

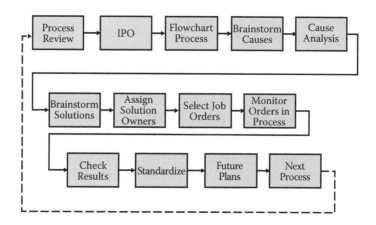

FIGURE 2.7
Error-proof process. (From Robert Baird.)

target. With the team formed, the next step was to receive training on error-proofing techniques. Every day for six months we focused 100% of our efforts and time on achieving the target of 99.9% process yield. We would select smaller work orders so we could prepare each process and run the order within three days. Please see Figure 2.7 for the process we followed.

As you can see, we used only simple Lean Six Sigma tools. The Input, Process, Output (IPO) tool was one of the main tools used to identify and characterize the key inputs of the process. These key inputs were then set to their optimal settings to obtain the output of 100% yield. Optimal inputs also included clean materials and machines and accurate work order information. The results, recorded in Table 2.4, were outstanding when you consider each process step ran at a minimum throughput of

TABLE 2.4

Yield by Operation

Order				Yield by Operation					
	Process 1	Process 2	Process 3	Process 4	Process 5	Process 6	Process 7	Process 8	Process 9
1	100%			99.93%	N/A	100%	100%	100%	99.97%
2	100%	100%	100%	100%	N/A	100%	100%	100%	100%
3	*Did not monitor job*	*Did not monitor job*	*Did not monitor job*	99.94%	N/A	100%	100%	100%	100%
4	*Did not monitor job*	*Did not monitor job*	*Did not monitor job*	99.98%	N/A	100%	100%	100%	100%
5	*Did not monitor job*	*Did not monitor job*	*Did not monitor job*		N/A	100%	100%	100%	100%
6	99.75%	99.81%	99.87%	99.40%	N/A	99.16%	99.98%	100%	100%
7				95.71%	N/A	100%	100%	100%	100%
8	99.90%	98.93%	99.87%	97.97%	N/A	99.82%	99.90%	100%	100%
9	98.56%	96.43%	99.62%	100%	N/A	99.70%	98.98%	N/A	100%
10	99.62%	98.93%	99.62%	100%	N/A	99.30%	99.95%	N/A	99.90%
11	96.25%	97.75%	98.62%	100%	N/A			*Did not monitor job*	
12	100%	96.06%	98.00%	100%	N/A	100%	99.90%	N/A	100%
13	97.98%	97.75%	99.31%	100%	N/A	100%	99.63%	100%	100%
14	95.96%	100%	98.62%	100%	96.92%	N/A	N/A	N/A	N/A
15	99.45%	100%	98.31%	99.81%	N/A	100%	99.85%	100%	100%
16	100%	100%	98.98%	100%	N/A	99.96%	100%	N/A	
17	98.00%	97.68%	93.81%		N/A	100%	99.90%	N/A	99.96%
18	99.9%	99.44%	98.31%	99.94%	N/A		*Did not monitor job*	N/A	
19	100%	97.62%	93.82%		N/A		*Did not monitor job*	N/A	
20	99.81%	97.94%	99.06%	99.16%	N/A		*Did not monitor job*	N/A	
21	99.93%	87.73%	100%	100%	N/A	100%	100%	N/A	99.90%
22	99.44%	99.19%	96.50%	100%	99.62%	99.44%	99.94%	N/A	99.98%

3,000 units/hr. Any process below 99.9% was recorded in red. As a note, before starting this effort no process step had a yield higher than 95.1%.

SELF-DIRECTED TEAMS PREVENT THE WEAKEST LINK

A process is only as strong as its weakest link. How many times have we heard this? What we need to answer is how we provide a more homogeneous process where all of the key process steps are strong and add customer value. In a traditional organization of "command and control" there will always be a weak link. This is because there are not enough improvement resources to ensure an equality or variation reduction for each of the key process steps. The answer to a homogeneous process is self-directed teams, where each key process step is owned by a skilled and empowered team. Ownership involves two key achievements: accountability of key business targets and the right skills to achieve pride in team achievement. Accountability and achievement are daily—not yearly, quarterly, or monthly. The two key achievements do not come simply by direction from management (command); they are the responsibility of management but not by "command and control." Management must practice a well-designed Standard Work for Managers and STP on a daily basis. To ensure the culture and fast-paced results this must happen daily. They must also understand that mistakes are going to be made and not to be punitive when they do. This is because trust is gained when not only mistakes are made but also learning occurs. Only with trust can we achieve an empowered self-directed team organization. Momentum is the key and cannot be gained by infrequent visits to the Gemba. Assigning process engineers or the quality department to process improvements takes too long and makes it difficult to maintain momentum. Self-directed teams who work at the process step 100% of the time are able to discover the key points and inputs that are required for standard work improvement.

With development of their continuous improvement skills, self-directed teams will find and resolve the process functions that cause the system to be weak. Their process scope is much more focused, so they are soon able to become experts and make sure their part of the process is functioning according to requirements. They are able to quickly bring problems to the surface and start to work on resolving them. I think most people know that process variation takes a very long time to reduce. This is because the problems are well below the surface and are not the problems that have already been addressed with process improvement efforts in the past. They are problems like information errors on work orders, something that stops functioning because of dirt buildup, a sensor that has stopped working, a sensor slightly out of adjustment, a wrong key process specification, a key transfer arm with a small bend, a scratch from something that touched the product, long shift change, poor job training, and so forth. These problems are very difficult for the process engineer or quality manager to realize with their infrequent visits by. Self-directed teams ensure their link is robust.

PROCESS IMPROVEMENT WITHIN THE OFFICE

The Toyota Production System (TPS) was developed between 1948 and 1975 by Taiichi Ohno, Shigeo Shingo, and Eiji Toyoda. TPS organizes manufacturing and logistics for the automotive manufacturing process. It was originally called Just-in-Time Production, and this is probably a more suitable phrase when thinking of process improvement within the processes of office work like human resources, sales and marketing, logistics, supply chain, customer service, work orders, invoicing, and the like. Each of these processes delivers either a product or service that our internal or external customers demand to have on time. To accomplish this objective consistently, the process must, as the name implies, deliver just in time. Not being able to provide just in time is a clear indicator of process waste. This is the basis and approach of TPS, elimination of process waste.

The first realization for people to improve an office process is just that—a process. Customer service receives a customer call (Step 1), information is passed to the planner (Step 2), ERP system is updated (Step 3), inventories are checked (Step 4), if low then needed inventories are ordered (Step 5), customer ship date determined (Step 6), and then customer service communicates to the customer the proposed ship date (Step 7). Of course there are more steps than this outline, but you get the idea that this is a clear process with many process steps. In fact, some office processes can have many more steps than the actual making of the product or service. And because there are typically many process steps, there is process waste, and as mentioned this is the basis of the TPS methodology: elimination of waste. Many of the proven tools used since 1975 in an automotive production system apply with office processes: improvement tools like self-directed teams, value stream analysis, cells, 5S, Standard Work, TWI, WIP reduction, Kanban, and Kaizen. In my experience applications of these tools bring outstanding results like 80% reduction in process cycle time and 60% increase in productivity. This is because the low-hanging fruit remains within these processes.

I include Lean office techniques as part of process improvement. Here are the reasons to improve office efficiency:

Increase productivity and quality of the product or service process (because the admins are inputs)

Support workers (experts) must be able to visit the actual value stream frequently

Management Steering Team, coaching, project support, 5S, and teaching are all applicable

Provide world-class lead time for our customers

Increase productivity of support workers by removing waste

Vision alignment

Improve morale and better communications

Remove silos

People who typically work within an office environment stay within this environment throughout the day. By this I mean they very seldom go to the Gemba. This is a huge problem because these people build tribes, a culture that is not conducive to world-class inputs for the value stream. Yes, they are inputs, but I have seen where they will make the value stream wait for their services because of the tribal culture they build by staying in the office. They must change to understanding that their internal customers are located at the value stream of producing products or services. There is no hierarchy in a world-class organization—only internal suppliers and customers all interdependent on striving to delight the external customer.

Eight Office Wastes

Overproduction—Too many signature levels, too many e-mails, ineffective meetings, more information than the customer needs, more information than the next process needs, creating reports no one reads, making extra copies

Transportation—Retrieving or storing files, carrying documents to and from shared equipment, taking files to another person, going to get signatures

Motion—Searching for files, extra clicks or keystrokes, clearing away files on the desk, gathering information, looking through manuals and catalogs, handling paperwork

Waiting—Waiting for meeting participants, for faxes or a copy machine, for the system to come back up, for a customer response, for a handed-off file to come back

Unnecessary processing—Meeting participants that are not required, creating reports, repeated manual entry of data, use of outdated standard forms or inappropriate software

Inventory—Files waiting to be worked on, open projects, too many office supplies, e-mails waiting to be read, unused records in the database

Defects—Data entry errors, pricing errors, missing information, missed specifications, lost records

Nonutilized talents—Probably the most significant waste; people not developed to their full potential

e-Mail Waste

Another rapidly growing waste is e-mail. The most important tasks should be completed when arriving at work in the morning. Unfortunately, we all open our e-mail the first thing in the morning and waste too much time on non-value added tasks, time that could be spent in the Gemba.

In a Microsoft Office survey (March 15, 2005), respondents said they are receiving an average of 42 e-mail messages per day. Translating that number into your most valuable commodity—time—workers are spending too much of their day managing e-mail, a number expected to increase not decrease. e-Mail wastes are as follows:

Colleague spam—that is, too many people are indiscriminately clicking the Reply All button or copying too many people on trivial messages, like inviting 100 colleagues to partake of brownies in the kitchen.

A good chunk of today's e-mails are also coming from brand-new sources, like social- and business-networking sites like Facebook and LinkedIn or text messages forwarded from cell phones.

The real reason why workers are wasting away their days in their inbox is that most of the e-mail is worthless.

Practicing Lean e-Mail

1. No internal e-mails within the building or the floor you work on.
   a. Go visit the person or call them on the phone.
      i. Communication improves.
      ii. Errors are reduced through misinterpretation.
2. When you send an e-mail to a group of people, put the recipients in the BCC field. That prevents them from clicking Reply All. (If you want to show who was on the list, put their names in the body of the e-mail.)
3. Let people know if a reply is required.
4. Pick up the phone.
5. e-Mail length:
   a. If you have to write more than two sentences, call them; if a group, organize a conference call.
   b. Single message—It starts with the subject line to describe the message, but restrict the e-mail to only one message or idea.
   c. Do not summarize a meeting in an e-mail; use SharePoint or other document sharing software.
   d. Do not summarize a call with a customer or other important stakeholders; this merits a call.
6. If you will be out of the office for an entire day or longer, turn on your Out of Office assistant with a message indicating when you will return and who can be contacted on your behalf.
7. Attached files should be limited in number and in total size. Consider placing a file on a common drive location and providing a shortcut link.
8. Store e-mails as files in your main directory and not within your e-mail software so that files of all types (e-mails, documents, spread-sheets) for one project are kept together and your e-mail works faster.
9. Set a target for the maximum number of e-mails remaining in your inbox—for example, no more than one full screen.
10. If you originate an e-mail message or memo, you are expected to keep the original on file for future reference.
11. Have an effective Management Steering Team in place.
    a. Facilitates communication between departments because they meet frequently, preventing the need for e-mail communication.
    b. Management Steering Team can set e-mail rules.
12. If you need to determine the best way to do something, hold a meeting.

13. Using e-mail software tools can lead you to being more effective and prepared when attending meetings. When accepting invitations to meetings, set a 15 minute reminder and use this time to prepare. Add 15 minutes to the meeting end time, and use this time to complete or plan meeting actions.
14. Keep your IT and security department abreast of any new spam getting through.

Some Time Management Tips

The most obvious way to manage your time is to schedule. Schedule monthly, schedule weekly, and schedule daily. Write down everything you need to get done. Now!

The moment you know you'll need to get something done, put it in your schedule—even if it's months away. Procrastination is the greatest enemy of time management.

Assign project deadlines two days in advance; this way you're always on time even when new projects are assigned.

Make weekly plans the Friday before and daily plans the previous afternoon. Try to do all of the important projects first thing in the morning so that they are done before you become overwhelmed with basic office administration and interruptions. Remember, this is not e-mail.

If you need to alter your schedule, get back on track ASAP!

Start an idea book and keep it handy. Later, when you have more time, you can give your ideas a bit more thought and rewrite them into effective plans of action.

Keep your desktop and files tidy. By removing the clutter from the surface of your desk, you're removing distractions.

Reward yourself for a job well done every time you finish a project.

To contact people who simply need a yes or no response, call them back when you know they won't be around.

Free yourself from the usual "How's the weather over there…?"

Move all of your less important e-mails from your Inbox to an Unread e-mail file until you either have a bit more time or just need a quick pick-me-up.

Regardless of how swamped you are, never deprive yourself of a lunch break, even if only 15 minutes. You may not feel hungry, but your body and mind need food to continue functioning at peak levels.

Effective Meeting Tips

Start on time.

Distribute a meeting agenda 2–3 days before the meeting.

Your agenda is the time management document for the meeting.

Assign a task for each participant.

If you cannot think of a task for someone, they probably do not need to attend. Each task is a line item on the agenda.

The meeting leader must prepare the room or conference call.

Ensure that—

Overhead projector is checked and running.

Conference call line is open.

Coffee and water are ready.

If meeting is consistently late, measure time wasted and why.

Invite the right people.

Consensus is difficult to achieve with groups of seven or more people.

Minutes of the meeting can be provided to people who would like to know the results.

Brainstorm to facilitate ideas when you're stuck.

Brainstorming rules must be stated and followed.

Kill the PowerPoint.

Limit it to training.

End on time.

Agenda includes time for meeting summaries and repeat of key points.

Each agenda line item is managed for time.

Put unfinished items on the next meeting agenda.

Conclude with a summary of decisions and action items.

There is a personal responsibility to be on time and prepared.

Do not book back-to-back meetings.

Use the meeting reminder time to prepare for the meeting.

A meeting agenda must be sent along with the meeting request.

Follow the subjects, do not wander, and stay within the planned subject times.

Meeting objectives should be stated and clear for everyone:

"By the end of the meeting I want the group to…"

Objectives are on the meeting agenda.

Confirm and confirm again.

Remind people the day before.

Assign actions.

Don't finish any discussion in the meeting without deciding how to act on it.

Follow up after the meeting.

Learn and develop your skills for effective meetings!

Provide structure: who will take notes, who will facilitate.

Lean Office 5S Tips

Sort

Sort through the Gemba walk with the Management Steering Team to red-tag items to be removed and managed

Review common areas

Sort copier, fax, conference rooms, visitor offices, coffee area

Determine if items need labeling, containers, or to be removed

Sort common folders with IT department

Sort any files that have not been used for over a year; locate them in a red-tag folder

Set a policy of clearing your desk at the end of day

Straighten

Label all common areas

Have equipment operating instructions

Have phone lists

Provide name and phone number of who is responsible for repairs of office equipment

Have Kanbans for office supplies (Kanban cards are faxed back to the supplier or the supplier actually monitors the levels)

Set a rule for nothing stored on top of cabinets

Provide restrooms with everything automatic (flush, turn on water, paper towel, or hand dryers)

Sweep

Clean daily

At the end of a meeting put everything back in its place

Standardize

Have common areas organized and labeled for general use

Have international signage for restrooms, exits, lifts, caution, etc.

Ensure that everyone knows exactly their responsibilities for adhering to the first 3S's

Sustain

Determine a 5S checklist and complete it monthly

Make the checklist visual

Include new ideas as part of the checklist

Include the sixth S—Security

Identify and address risks within the office environment.

All PC computers to be secured with cabling, unattended PCs to be secured with password, confidential documents stored, lock secure information rooms, etc.

Summary

With all of the Lean office tips provided above, I have determined a productivity gain of 3.97 hours per person per day. This might seem high to some people, even the most productive person. Please take the time to do this little exercise for the week:

1. Determine the average per person per day your organization wastes in each of the categories from Figure 2.8. Sum these categories.
2. Take the sum from step 1 and subtract 3.97 hours.
3. Calculate the percentage gain in productivity per person.

Example:

1. Average per person per day wasted hours sum to 5 hours
2. From 5 hours of waste subtract 3.97 hours saved by practicing Lean office = 1.03 hrs
3. Productivity gain = (1.03 – 5)/5 = –79.4%

FIGURE 2.8

Total office 5S time savings. (From Robert Baird.)

So what can you do when you remove the non-value added activities? Turn the time into value-added: Go to the Gemba and apply the techniques of STP and learn.

GOVERNMENT PROCESSES AND LEAN TECHNIQUES FOR IMPROVING FLOW

How can government value streams benefit from application of Lean tools and techniques? There are many ways, especially in the improvement of process flow. Takt time, FIFO, and Pacemaker are very applicable Lean tools that ensure an effective pace of processing. They are explained in the next paragraphs.

The Use of Takt Time and FIFO

If the process is services only, like people retiring, then the administration process is a FIFO system (First In, First Out) from the request point back to the customer. The first step is to look at all of the individual process steps and get an accurate measure of processing time in seconds/request or transaction. Plot this time and all process steps as a bar graph. Once you know this then look at your customer demand and determine Takt time (available seconds of work per day divided by units required per day). Plot this time on the same graph, secondary axis (it will be a straight line). This will provide a visual of which process you will need to focus your improvements on first to improve flow. For the processes that cannot meet this Takt time you will need to add resources or look at ways to improve the cycle time. This cycle time improvement is accomplished through the removal of wasteful steps. I found in administration processes it is mostly one of these wastes: transportation (next process step is too far away), waste of ability (people waiting for validation), and overproduction (faster processes building WIP, like completed forms faster than the next step can process them). You must build your complete administrative system to about 80% of the highest request period (lowest Takt time). To do this you must identify and remove wasteful steps at every point in the process.

Another good tool in this analysis is called the Pipeline Map. Plot your actual daily output (for a month) of each process step, and then plot what you should be capable of for daily output for each process step, all on the same chart. This will give you a visual of which processes are being over-utilized (bottleneck) and which processes are being underutilized. Then go to the Gemba and observe why this is happening. Again it will help you to focus your improvement resources. Another method to increase your overall capacity is to look at cross-training your people to complete tasks in slower process steps. This will help in the total number of people you will need. I know people will say you are not allowed to do this and will question why. Until you remove the process wastes, you will have to hire temporary people for the peaks, but you will begin to need fewer of these temporary people as your process improves. Use the cycle time (minutes per request) as your main indicator—a single focus. Have the people in the process responsible to manually update this chart, with the target of getting below the customer Takt time at peak demand. Visually manage this chart located at every process step so managers can support the effort. Then the overall project improvement team can produce a weekly chart of all the process steps compared with Takt time (updating chart 1). Now this is important: You need the people in the process to help improve the process steps, so you need to train them on how they can contribute with a problem-solving methodology and give them time to work on the problems they find, say one hour per week. Last, look at office cells; put the people in each of the process steps close together and in a U shape. You will be amazed at how the overall process cycle time will improve by just doing this. Then in the future if you want to take advantage of the benefits of self-directed teams, you have a natural grouping.

The Pacemaker Process

The Pacemaker Process sets the pace (processing rate for every step in the process) for all of the upstream process steps. It is the only process step that is scheduled, so if you have fluctuating services at the Pacemaker Process, then all of the upstream steps must adjust their capacity to match. For example, in a certain government administration process offering services to citizens, such as passports, the Pacemaker would be the office where people apply or the website for application. The upstream processes like application processing (and steps within this process), Government Printing Office (GPO), and supplier would all be connected through a supermarket pull system. The supermarket holds the WIP of small, large, foreign diplomat, and government employee passports. These supermarkets are sized according to 85% of the requests (highs and lows). The Pacemaker (the first process in this case, and typically the process closest to the customer) sends a signal (withdrawal request) back to where the supermarket is, upstream, and that process then produces to the supermarket to replace what was withdrawn. The upstream process is not allowed to provide more passports than were withdrawn. This prevents waste of overproduction by faster processes upstream and thus an increase in lead time because of too much WIP.

REFERENCES

Goldratt, Eliyahu M. 1984, 2012. *The Goal.* Great Barrington, MA: North River Press.
Microsoft News Center. 2005. Survey Finds Workers Average Only Three Productive Days per Week. (March 15).

3

Component 3: Organization Structure for Sustainment

In order to support, teach, and promote (STP) a fast-paced, continuous improvement, a strong organizational structure must be in place, a team-based organization. Ideas, information, and solutions to issues must be able to flow unimpaired. In most organizations it is completely the opposite, starting with policies that minimize the contributions that front-line workers can make: jobs are designed to be idiot-proof. Technology is used largely for monitoring and control. Pay is poor. Training is minimal. Performance expectations are abysmally low. There is no structure of support; "us and them" thrives. Successful companies are developing a service-driven logic where there is an interdependent and responsibility system of internal supplier and internal customer all along the value chain. This system provides a natural support system and a strong connection to the external customer, something missing in today's value stream. Every organization has its experts and unsung heroes, but their knowledge and abilities are like floating bubbles within the value streams; it takes too long to transfer. Employees have many ideas; the problem is they go unheard due to the hierarchical organizational structure of command and control. When the structure is less vertical and consists of empowered employees, ideas are put into place without one, two, or three other levels having to make decisions, which subsequently slows and suppresses the implementation of the ideas. The idea behind a resilient organizational structure is mainly support—the support required for people to feel comfortable in contributing toward meeting the strategy, the support in feeling comfortable to take a risk, the support in feeling comfortable in using newly acquired skills, and the support when a mistake is made. Self-directed teams provide the main organizational structure, but structures also need to be in place for the implementation of key programs like Lean Sigma and TWI (Training Within Industry).

ORGANIZATIONAL CULTURE FOR ENGAGEMENT AND CREATIVITY

Every company has its own personality or culture. It is the way business is done, the way people interact, the way decisions are made, how promotions are determined, the methods used for improvement, and how information and knowledge are used, communicated, and valued. Many organizations do not specifically think about the importance of their culture. The thinking is more toward how to best structure for deployment of their strategy. This is one of the reasons for the traditional hierarchal and layered organizations. The irony is that this organizational structure comes with a high cost. For some reason cost is always the driver for improvement—an output that of course needs inputs to be improved, and organizational culture is a key input for not only cost but overall productivity. Providing an organizational culture of engagement and creativity will provide fast-paced results in productivity and quality. The organization is one of empowerment and a coordinated movement toward making the organization successful. The organization is much leaner because span of control for the managers is significantly increased.

The Japanese term *Genchi genbutsu* refers to "getting your hands dirty." It is practiced by even the top leaders of organizations. Everyone is expected to go to the Gemba to understand where improvements are needed. With visual value streams and daily visits to the Gemba, management can make real-time decisions, preventing potential errors and wasted tasks. Improvement and strategy-related projects are completed faster with fewer resources. This is the culture of companies successfully practicing Lean Sigma. Organizational layers do not work well for engagement; they were made for compliance and for knowledge protection.

I travel to many different countries, and when I start to discuss programs such as autonomous or self-directed teams, organizational management gets very defensive. The common statement I get everywhere is, "You are from America and you do not understand the country culture," and this will never work here. I find this statement very interesting because these are very intelligent and educated managers, but for some reason they do not see the connection of how *they* are the culture setters within the four walls of their organization. Sure, some of their country's culture is within their four walls, but leaders must establish the dominant culture of how business is conducted.

What are the steps to ensure an engaging and creative culture?

1. Develop a structure for knowledge sharing—This starts by establishing a Management Steering Team for strategy deployment with the objective of engaging 100% of the people within the organization. Most department managers are automatically members. With the Management Steering Team established, it will start to become apparent that departments need to work much more closely together and important organizational decisions are more by consensus (see Chapter 1 for more about the Management Steering Team).

2. Implement self-directed teams—This is the key. You do not realize how many superstars you have within your organization until you start self-directed teams; more on this in the section titled Self-Directed Teams: A Culture of Engagement, later in this chapter.

3. Use STP during Gemba walks—Here are the main methods of STP; from there you can easily understand how these methods contribute to engaging and adopting this culture:

   a. Support—Used to remove roadblocks and reinforce expected culture. Roadblock removal can be providing needed resources, facilitating a decision needed by another department or team, listening and acting on a suggestion for improvement, confirming certain process requirements, and so forth. Reinforcing expected culture is personally demonstrating behaviors and even using the right language. Allowing mistakes to be made will build trust and allow people to be more confident in their improvement efforts.

   b. Teach—Anytime you recognize that someone does not understand a program, how to resolve conflict, policy, safety procedure, quality control, key points of the strategy, Lean Sigma tool, and the like, you must become a teacher. All managers must become very good teachers. People will always look up to people they learn from and are quicker to follow a competent teacher. I talked about mentorship in the component of Leadership and Mentoring; teaching is the most powerful element.

   c. Promote—Promotion during strategy deployment is key in people clearly knowing the strategy and in understanding how they contribute. In most organizations, only the people who are privileged to be there during strategy development (and this is questionable) know and understand the reasons for the strategy. After strategy development, a general communication meeting is held and objectives are handed down, expecting that everyone now understands both the strategy and how to contribute to the strategy. This is a grave assumption. During promotion, start by asking two simple questions at random: "What is our strategy?" and "How do you contribute?" When people do not know, use teaching methods to promote the clear understanding for 100% of your people; the goal is 100% of your people. The leaders are also responsible for promoting the expected culture by identifying behavioral patterns and methods of accomplishing the strategy. If they are different from the expected culture, then teach; if they are what is expected, then support.

ORGANIZATIONAL STRUCTURE SUPPRESSING IMPROVEMENT

How many of you have started a Lean Sigma effort in one of the organization's vertical structures, gained some extraordinary results from people engagement, moved on to another project, went back to follow up on the successful effort, and found the improvement solutions have almost disappeared? Most people will determine the root causes to be lack of leadership, lack of empowerment, and lack of motivation. But are these not symptoms and the actual root cause the organization's structure? The structure of an organization will determine how the organization will operate and perform. Most organizations are hierarchically structured with multiple levels. The chain of command looks like a pyramid where the large base of workers is directly supervised by a smaller level above them and they in turn are controlled by an even smaller level until you reach the level of CEO. This type of structure is slow to react to new opportunities, resistant to innovation, and slow to react to today's dynamic business conditions.

The traditionally structured organization relies on top management to make the decisions. In a team-based organization all employees participate in making decisions. They are empowered to make decisions related to their scope of the horizontal process. They know the abilities of their internal suppliers and internal customers, and the horizontal value-added process is interdependent so that speed and quality improve. Employees feel they are part of the total organization because it is clear how their efforts are contributing to the overall business success. This contribution provides individual value, which in turn provides a high level of motivation. So when there is a successful improvement effort it is more likely to sustain as the self-directed teams directly feel responsible for the outcome of their part of the process. In a hierarchal structure, the decision maker is too far away from where this improvement was discovered and will most likely make decisions based on what worked before.

If you have felt the frustration of diminishing results from a successful improvement project, think about your organizational structure. Is it actually conducive to a fast pace and sustainment of business results?

SELF-DIRECTED TEAM ORGANIZATION

Together extraordinary people can achieve extraordinary results

—**Becka Schoettle**

Shop Floor Self-Directed Team Definition

A shop floor self-directed team is a group of operators facilitated by mainly the shop floor supervisor. The group can be defined by the operators responsible for a cell and/or shift, operators responsible for a grouping of machine type, operators responsible for a function like shipping and receiving, operators responsible for a product line process (value stream), and so forth. The team size can be four to six team members. I caution when exceeding six team members, as it can become very difficult to manage and develop the team with this large diversity of personalities and ability. The team also will start to develop teams within a team because of personalities. This can result in severe conflict, which can be very difficult to unroot. If this has to happen, you must have a very strong and experienced team facilitator.

The main objectives for establishing shop floor self-directed teams are to provide fast-paced problem solving, provide sustainable results, gain productivity, and improve quality.

Self-directed work teams represent an approach to organizational design that goes beyond ad hoc problem-solving teams. Members operate with a high degree of trust, accountability, and interdependence. Self-directed team members share authority and responsibility for self-management. Members create synergy with a strong sense of mutual commitment. Members help one another; they help other team members realize their true potential and above. These teams are natural work groups that work together to perform a function or produce a product or service. They not only do the work but also take on the management of that work—a function formerly performed by supervisors and managers. This allows managers to teach, coach, develop, and facilitate rather than simply direct and control. These teams are empowered to take corrective actions to resolve day-to-day problems. They also have direct access to information that allows them to plan, control, and improve their operations. In short, employees that comprise work teams manage themselves. With a cell layout, each self-directed team owns a part or all of a value-stream. This makes it clear for each individual where their main responsibilities are required.

Businessweek recently reported that self-directed work teams are, on average, 30% to 50% more productive than their conventional counterparts (May, n.d.). The following are some examples of organizations that attribute major productivity results to the advantages of self-directed work teams:

- AT&T—Increased the quality of its operator service by 12%.
- FedEx—Cut service errors by 13%.
- Johnson & Johnson—Achieved inventory reductions of $6 million.
- Shenandoah Life Insurance—Cut staffing needs, saving $200,000 per year, while handling a 33% greater volume of work.
- 3M's Hutchinson facility—Increased production gains by 300%.

Why is this concept of self-directed teams growing? Leaders of organizations are now focusing on how to engage their employees in order to create a faster pace toward meeting targets. To do this leaders are realizing the organizational structure must change and self-directed teams are part of this structure. Self-directed work teams can result in

- Improved quality, productivity, and service
- Greater flexibility
- Reduced operating costs
- Faster response to technological change

- Fewer, simpler job classifications
- Better response to workers' values
- Increased employee commitment to the organization
- Ability to attract and retain the best people

Targets unachievable in a traditional and hierarchal organization will become a realization with a self-directed team organization. The deployment of self-directed teams must be well planned, including expected business targets from this new organizational structure, emotional impact on the people going into teams and the people supporting them, job description changes, and new skills to obtain.

Teams … I can hardly wait for the Management Steering Team's announcement of this new and wonderful way of working. The workforce will be jumping for joy when they realize that teams will give them more autonomy, provide them with more problem-solving ability, provide them with more authority over their process, provide them with a voice, and enable them to make the most of their own decisions. In reality, the zealous team in the workforce may not be there; and in fact, certain parts of the workforce, namely people paid by the hour, might be hiding some intense fear of this change. Why? First, change always creates anxiety, and for years these people have relied on their immediate supervisor for decisions to be made, especially when there were some tough problems facing them. They have an unspoken dependency and comfort with their supervisor. What you appraised as your better hourly people might become your principal opponents of the team concept. Don't forget that in the past these more productive people probably received special treatment and rights. In the team environment, they can see that the team purpose and results are now the focus. The team gets the reward and recognition—not a few individuals. They might have concerns and questions about having to work more closely with people they appraised as not pulling their own weight, why they have to change when they thought they were doing OK, that added team meetings will take away from what management really wants, which is production!

Teams do not need supervision; what they need are systems, which help workers understand their direct impact on the business. To this end, the role of the team facilitator is to teach, train, and coach team members in the many skills required to become self-managed (empowered). This begs the question Why are there supervisors? The supervisor function was developed to gain control over the line workers because the organization did not trust them, nor would the organization train them in how to work.

It simply seemed easier to hire someone who already had much of the technical knowledge to oversee the work.

In the transition to self-directed teams, little will be left of the old control paradigm of management. The supervisor will also need to transition to a team facilitator role, where they will need further skills to work with and support self-directed teams.

Implementing Shop Floor Self-Directed Teams

Shop floor teams can be started at any point of the Lean Sigma implementation, but the sooner you start, the faster you can start realizing the benefits of 100% of your plant population contributing to the strategy. These teams must be started after the cells have gone into place because the new layout will define their scope of responsibility, which will help create process ownership.

Sustainable world-class results are impossible without shop floor self-directed teams. The shop floor employees have inherent process knowledge that is difficult to obtain by management and engineering positions. Finally, if you have a culture of resistance to change, start self-directed teams; they will become your strongest supporters.

Shop Floor Self-Directed Team Process and Requirements

In a sports team, everyone has a position to play; and when each person is playing their role effectively, success happens. This will also create interdependencies, again facilitating success. Like in sports, self-directed teams will require team structure both for conducting effective problem-solving meetings and for day-to-day production requirements. Before the big game and throughout the existence of a sports team, extensive training and coaching is provided. Self-directed teams will also require training to get started and continuously improve. Also throughout their existence they will continue to receive training to continuously gain required skills and knowledge. The ongoing training will become part of their responsibilities during planning sessions. The coaching will come from their team facilitator and the Management Steering Team during their daily Gemba walks.

Continuous Improvement Responsibilities

Self-directed teams must be structured with a team facilitator, team leader, team notetaker, and team timekeeper. These are fundamental positions required to conduct effective problem-solving meetings. All of these positions require training for the complete team as the roles are rotated (except for the facilitator). The team facilitator is typically the supervisor (who is transitioning to the new role in supporting the self-directed teams) who has received the required training in

- Team facilitation
- Resolving conflict
- Problem solving
- Conducting effective meetings
- Mentoring

The team also requires a name as determined by the team themselves. This provides an identity for the team, and once named, all future reference to this team must be with their name.

It is mandatory that the team meet for at least one hour per week for their continuous improvement activities. There is no excuse for canceling a meeting; they could move it to a different day in the current week, but they must meet at least one hour per week. We found as the teams mature they will actually require more than one hour per week because they sometimes are very close to a groundbreaking solution. One of our more mature teams who had reached high-performance status actually took their complete shift off to volunteer at a local grade school. They never missed and in fact in most cases exceeded their key metric targets. When telling this story to plant managers, I usually elaborate more because when managers first hear the requirement of one hour per week, their paradigm does not allow them to think of it as an investment toward quality and productivity improvements. Their first thought is of chaos and loss of command and control. Of course they will soon realize the benefits and their paradigm will change.

In the beginning, the meeting agenda must accomplish the following:

- Establish who will take the various team positions (completed by voting or volunteers)
- Determine the day and time the team meeting will be held
- Decide on the team name
- Determine team rules and norms that everyone agrees to

Training is needed to get the teams started and to answer some of the questions each person will be asking before going into teams: What is expected of me? How will we function? What if the others do not like my ideas? What will happen if there is conflict? Will I need new skills, and if I do, will I get more money? How does a team make decisions? What decisions will the team be responsible for? What will my supervisor do? What's in it for me? All of these questions must be answered during the training.

Once these team fundamentals are determined, they can start working on their first process problem to solve. The Management Steering Team needs to set the completion time for all of the team's projects; one per month is ideal. Providing more time will only lead to projects that are too complex at this time.

Process problems are usually selected by process data (Pareto chart of current problems). The teams learn quickly that any improvement projects selected must clearly contribute toward the single-focus strategy. It is important to start with simple problems because in the beginning they must become familiar with the problem-solving methodology. The scope of the problem to solve must also be small. For example, focus the problem solving on one machine type or one part of the process. Do not select or assign a problem to solve all defects related to a certain machine group. Select one machine and the most frequent defect within that machine group and keep their focus there. It is also important to select a problem the team can solve by themselves.

A problem-solving method must be selected for all teams to use (Define, Measure, Analyze, Improve, Control, or DMAIC; Plan-Do-Check-Act, or PDCA; 7-Step, etc.). This will provide a common problem-solving language.

Once the team has completed their project, they must present to the Management Steering Team. The presentation includes the content of each of the problem-solving steps, and usually a PowerPoint presentation is used. The Management Steering Team then has three objectives: provide comments on the problem-solving process, approve any resources required to permanently implement their solutions, and approve the project solutions. Finally, when the teams present their next project, they conclude the presentation with a demonstration that shows the results of the previous project are being sustained.

Team Process Improvement Metrics

It is obviously important to have metrics for the self-directed teams to enable them to realize their own progress and their contribution to the single-focus strategy. The key is to have metrics that are understandable and with a narrow scope. Self-directed teams will not relate to output metrics such as lead time, detects per million opportunities (DPMO), standard (STD) cost, Cpk, throughput, or even cost of poor quality. They can relate to and control input metrics like the number of units they scrap for a certain defect at a specific machine, level of work in progress (WIP), the time it takes to do a particular task, how many times they do not have a certain material or information available, how many times a hot stamp die has to be cleaned in a shift, how long their machine runs before it stops, how long shift change took, how many times they exceed a specification limit for a certain value-add process, and many more examples like these. All measures must have a connection to the strategy. They can relate to input metrics because it is what they can control and immediately see the impact of their efforts; it is what they do every day. These inputs obviously affect the output metrics I listed above, so the Management Steering Team must make this connection for the team; it will show the team how they contribute toward the overall strategy. As you can see, these metrics will have to be manually tracked; this will actually help them work together as a team in the data-gathering exercises. The targets can be set after a baseline is established.

With the understandable and narrowly focused metrics selected, the team needs to standardize a method of how they will review the metrics. This method will be daily and part of the management's tiered Gemba walk. There will be a standard three-sided visual management board prominently displayed within their process that can be used for daily review and updates. Each one of the metrics must be assigned to a team member with the responsibility to provide current and accurate data updates. The three-sided visual management board provides metrics for digging out process problems, current problem-solving projects, and sustaining achieved results

Self-Directed Team Decision Making

Decision making within a self-directed team will become part of the Team norms but it needs some debate. Will decisions be made democratically (by voting), by consensus, decided upon by the leader, or by a combination of all methods? Whatever method is decided, all team members must respect the decision. It is probably the most important team norm as it will prevent conflict within the team. My experience shows that the democratic method has worked the best and then some of the more mature teams have moved on to consensus. The team facilitator will have to be prepared to coach the team on the method and then to maintain the discipline of the method selected. From my experience it is also very difficult for all team members to accept a certain decision. This again is where the team facilitator or a mature team leader needs to immediately intervene with the team member not in full agreement. This must be done in private after the meeting. The first explanation is pointing out that this decision followed the method within the team norms and it will contribute to the team's improvement targets. It must be done immediately because it is related to conflict, and the longer you allow conflict to go on, the more difficult it is to remove.

Management Steering Team Establishes Standard Work for Team Reviews

Self-directed teams need to clearly feel that they are contributing to the strategy. This will be accomplished with the tiered Gemba walk. The first step is to set up the team's three-sided metric board with the following measurements on each side:

1. Pitch attainment to "dig out" the process problems so they can be understood and removed. The pitch you are attempting can start with hourly measurements of units produced. An hourly target is set, and when the target is not met the team's responsibility is to record the reason. What was the problem. The reasons are then added to a list, which becomes a Pareto. A matrix is also created to record when a pitch is not made within each hour of the day. This will provide a visual of which hour(s) of the day are the most problematic. For example, in the beginning it is usually most problematic at the starting or ending of a shift. The team can obviously do something about this problem because it is related to shift change activities.
   a. Pitch can start with units produced but it should be aligned with the single-focus strategy. If lead time is the single-focus strategy, use the team's process cycle time for the pitch. It will still be hourly and of course a target so they can determine when to record a reason for not meeting the hourly target.
   b. Continuing with process cycle time from point a., if the target process cycle time has been set to 3 hours and the process easily meets a cycle time below this target then the target has been set too high.
2. The second side of their board is their current problem solving. The team displays their current problem-solving project progress

(please see the section later in this chapter entitled, "Dynamic Problem-Solving Process").

3. The third side of the board is the metric showing the success of their last project. This is kept current for three months after the completion of the last project. It will provide the indicator of sustained success. Also on this side of the board is their TWI training matrix used to ensure that progress is being made toward standard work and the team's career path.

With visual management in place, the tiered Gemba walk becomes much more effective in using STP (supporting, teaching, and promoting). The visual management provided on the standard three-sided board will provide management with the pulse of the process, project progress, and establishment of standard work along with its improvement. All three of these measures will be aligned and have line of sight to the strategy. The importance of this visual management, because it is set up throughout the value streams, is the ability to provide improvements on a daily basis.

TEAM STABILITY

To keep the team stable, do not keep changing the team members. This is a huge mistake. Team members must get to know each other; how they work together becomes an effective dynamic. If you remember, the first U.S. "Dream Team" basketball team never did as well as expected even though they were a team of superstars. It is the same in the World Cup. These individuals do not have the opportunity to play together to become the most effective toward meeting the team objectives. Changing members frequently will never develop bonds and free-flowing information required for team identity. This identity results in individual value and individual security required for team pride. A team with pride will always be a strong contributor toward accomplishing the organization's strategy.

Standard Work Improvement Responsibility

Self-directed teams must have a process to challenge the current standard work. They must know this is part of their responsibility. This process to challenge is part of Training Within Industry (TWI). This is one of the reasons for the team facilitator becoming the senior trainer and also why we put a TWI structure into place. The senior trainer mentors the team on developing the key points. The TWI Key Points are what the team works on; they are looking for key points that have not been discovered or are developing the ones in the current standard work. When there is a problem, they first determine the root causes and then determine if the current standard work has sufficient key points to prevent the problem. If not, the key points are either added or improved to address the identified root causes. Next the job breakdowns are updated, and finally the team members are trained on these new key points. This key point development cycle is learned quickly and becomes a staple of the continuous improvement program. This method allows the self-directed team members to have ownership of many improvements. This ownership provides team pride.

Management Steering Team to Establish Team-Based Reward Systems

The traditional reward systems are focused on individuals. A self-directed team structure must have a combination of individual and team-based rewards. Here is what we used for team-based rewards:

1. The team is rewarded for achieving the process improvement metrics aligned with the single-focus strategy. This is accomplished by a bonus system.
2. Individuals are rewarded with a symbol (badge, card, ceremony, etc.) for achieving their career path levels. The team is rewarded by having *High-Performance Team* stenciled on their metric board for achieving their levels of empowerment.
3. The team is rewarded through a team appraisal system designed by them. Individuals are appraised on how their efforts contribute toward the success of the team.

Along with this direct rewarding system they are also encouraged to learn and share knowledge. It is also OK to learn from mistakes. It is very powerful when a team learns new key points and they, as a team, teach other teams. The Management Steering Team and the team facilitator must be careful that the team does not try to protect their competitive advantage. The organization is about continuous improvement of the complete process and not a competition of teams.

The Mission Is on the Single-Focus Strategy

Teams must understand that they are there to contribute to the single-focus strategy. This will provide individual value and motivation as they feel they are part of a higher purpose. It will also prevent conflict within the team with the understanding of a common mission.

Getting the Implementation Started

The implementation of the shop floor self-directed teams starts with a plantwide communication. The Management Steering Team needs to carefully select the content that will be communicated.

This message needs to have the following content:

- Who will be structured in teams and how the teams will be selected
- The reasons for going to a team structure
- That it is required to reach world-class results that will provide job security and profit
- How it enables everyone to contribute
- That training will be provided
- When it will start
- What it provides for them
- How it will provide a process for everyone to participate and contribute in the achievement of the business strategy
- How it will increase their skills
- How it will empower workers
- How it will work

After the message is communicated to everyone, team training needs to be scheduled immediately. The training will include how to work in a team, what the various roles will do, the stages teams go through, how to conduct effective meetings, and the benefits of teams. The facilitator training is a separate training session and will include how to facilitate a team, mentoring, how to keep a project on time, how to resolve conflict, how to develop the individuals in the team, how to progress the team to the High Performance Stage, how to support the team member career path, and how to hold effective meetings.

Once the training has been conducted, the teams can start holding the first of their continuous meetings. The team will establish their team structure, and then the first problem-solving topic will be selected. At this time the department managers assign their teams the yearly or half-year objectives. Remember to keep them understandable and with a narrow scope.

TEAM EMPOWERMENT

One of the objectives with the shop floor self-directed teams is to have the teams start making more and more of the daily decisions related to their process. The team empowerment steps enable high performance.

Below are some examples of team empowerment and time frames to accomplish them:

Month 1—Determining daily work schedules, do we have efficient supplies, what targets do we have to meet today, who will run what equipment/tasks, who do we have to cover for, how much time for continuous improvement activities

Month 1—Problem-solving process, who is responsible for recording progress (A3 updates), how much time to allocate, how to communicate, skill development progress, sustaining gains, TWI-JI Key Point development along with standard work development

Month 1—Team norm development

Month 3—Determining review and feedback process for accomplishing team goals, how we will record the progress, who is responsible, how much time to complete this process, what happens if we are behind, communication of results to management

Month 4—Vacation schedule setting, how to set, how to manage

Month 5—Quality control and assurance, how to ensure understanding of assurance and execution, each member's responsibility, how to improve, how to communicate to our internal suppliers and internal customers

Month 7—Overtime control, how to determine, what are the targets, how to prevent, what are the causes

Month 9—Autonomous maintenance, managing preventive maintenance (PM) schedule, development of standard work for maintenance tasks, value-added cleaning tasks

Month 11—Team appraisals, what is the process, what is the assessment criterion, how will we provide a rating

Month 12—Planning for necessary training, career path development

Month 14—Budget responsibility, what is the team responsible for, determining variance targets, review frequency

Empowerment is not an immediate given right; you cannot just say, "Now it is the team's responsibility to complete this task." You cannot provide the skill effectively by only providing training. This is a recipe for disaster. A careful plan needs to be identified for each task on how the *team* will acquire the ability and knowledge to successfully complete the task. This plan is the responsibility of the Management Steering Team. Once provided with the empowerment, measurement and observations are needed to ensure that the team is successful in managing the task. The plan must also include an updating of Standard Work for Managers, where the teams are supported and further coaching is provided during the daily Gemba walk. Some fine-tuning will probably be required. The Lean leader within your organization can provide empowerment training. You can also hire an outside firm for expertise you do not have in house, such as for team appraisals. However, as I said before, the planning and the implementation is the responsibility of the Management Steering Team, so they must become versed in these empowerment tasks. Without this knowledge, the supporting and coaching cannot take place, so be prepared.

To get our teams started on obtaining not only higher levels of decision making but on how to become high-performance teams, we had our Management Steering Team design levels of empowerment. We would visit other team organizations, and during these benchmarking efforts we would ask if they had high-performance teams. The common answer was yes. We would then ask how they determined high performance, and that is when we received a blank stare. You see, to most people, *high-performance team* is just a label, and we did not accept this. We wanted something that told us our teams had achieved a high level of performance. We determined that high performance was directly related to the level of decision making they were *capable* of and to positive results. Of course this meant the teams had to be empowered. Listed below are the various levels of empowerment we developed. The teams then strived to complete each of the line items within each level. The fourth level of empowerment is where they do not need their supervisor any longer. They have autonomy and a much higher level of skills.

- First level of empowerment:
  - Select team leader and other team management position (facilitator)
  - Review and redesign layout of work area
  - Establish functional job assignments for team members
  - Set performance goals aligned with single-focus strategy

- Establish ground rules (norms)
- Schedule workloads and assignments
- Start problem-solving method
- Second level of empowerment:
  - Develop organization and team mission aligned with single-focus strategy
  - Start cross-training and support training
  - Start tracking performance against performance goals
  - Conduct regular internal supplier, internal customer improvement meetings
  - Establish team self-discipline to enforce ground rules
  - Establish self-assessment process.
  - Start Autonomous Maintenance tasks in Standard Work
  - PM schedule outlined and adhered to
- Third level of empowerment:
  - Teams to schedule workload and normal hours of work
  - Make decisions on overtime
  - Practice frequent and direct feedback
  - Develop key points
  - Develop standard work
  - Conduct regular communications with other teams and management
  - Interview new team members from internal requests
  - Establish hiring guidelines for new team members with management
  - Conduct team member performance appraisals
- Fourth level of empowerment:
  - Team is making most of their own decisions
  - Team is able to provide customer tours of the complete process
  - Problem solving is autonomous (team facilitator no longer required)
  - Project results have proven sustainment
  - Team follows a process of improving standard work
  - Team takes part in hiring new team members
  - Team is responsible for some part of budget

With a structured empowerment system, the self-directed teams can work toward continuous improvement of their part of the business, but always contributing to the single-focus strategy. When achieving all four levels, they can proudly say they are a high-performance team. This we realized was very powerful in individual value and motivation. The dialogue between team members changed to how to improve to the next step, how can we do even better.

Self-Directed Team Member Career Path

Along with the empowerment skills required to become a high-performance team, other skills are required to continuously improve their part of the process, meet the day-to-day requirements, and work with and support teammates. The career path also provides each individual with a pathway toward becoming part of the management team. The career path is designed, developed, and supported by the Management Steering Team, who reviews and asks what skills the teams will need to contribute toward the single-focus strategy and provide them with individual value.

The Management Steering Team starts by reviewing the current job description and changing it to reflect the objectives and responsibilities of a self-directed team member. The next step is to determine the levels of the career path. We simply used the following:

- Team member trainee
- Team member
- Senior team member
- Team member specialist

The team member trainee is the entry level, so these trainees were required to get to a fundamental level of running one process step within the cell. This qualification required the person to achieve this level within three months, and if not, they were released from employment. The team member level was a "rite of passage"; it was a set of skills the team decided as requirements to be a member of the team. This meant they were able to effectively contribute to meeting the team's targets. The senior team member attained skills like TWI-JI (Job Instruction) trainer, led problem-solving projects, and was considered a team expert in at least one part of the process. The team member specialist attained skills enabling them to lead the team through the day-to-day work, assign daily work and improvements to make, communicate requests to the Management Steering Team, and take customers on plant tours. At this level they are the obvious selection when a management position comes open. The beauty of this is they are absolutely ready for management, so please also provide them with interpersonal skills.

Here is where the traditional manager struggles. Each career path level will now require a review of current salaries and will require an increase. This to me is obvious because, if the skills are selected correctly, these individuals will now be significantly more valuable in the organization's ability to meet, at a fast pace, the single-focus strategy.

Another important part of the career path is the validation process of the skills being paid for. We handled this by empowering the teams themselves to validate these skills. In the beginning it did require the teams working with and the support of human resources.

WAITING, THE GREATEST WASTE OF ALL

How many of you have considered and therefore measured waiting time as a symptom of low productivity and lost value-added time? Waiting is part of what is called the "hidden factory"; we see it every day but have a difficult time processing it in our minds. In both manufacturing and service organizations, it can be the number one waste. We wait in queues, for medical services (even the room is labeled the *waiting* room), for maintenance requirements, for government services, for food preparation, to get somewhere (average person waits 3 hours per month in an airport and 15 minutes a day waiting in traffic), for information, for legal services, and so forth. In our personal lives we have learned to adapt to this wasted time by working on our mobile devices, catching up on necessary reading, and following up with phone calls. We attempt to turn it into value-added time.

What are the main categories for waiting in business?

1. Unbalanced process—One process step has a longer cycle time than the one it is an input for. For example, process step 1 has a cycle time of 120 seconds and process step 2 has a cycle time of 90 seconds. This not only will have an impact on productivity but also will increase lead time because of WIP increase. A process can also become artificially unbalanced when people are out sick or on vacation.

2. Running out of raw materials—This can be actual materials needed for the bill of materials, or in services it could be the required forms or even customers to serve.
3. Waiting for information—This waste is usually hidden from the leaders of the organization but the people working in the process know this waste very well.
4. Quality issues—The previous process has to rework or is waiting for decisions from another department like quality or engineering.
5. Equipment maintenance—A machine goes down; this includes computers needed in the previous step.

How to resolve waiting and improve productivity? A self-directed team organization can effectively resolve many of the causes listed above:

1. Unbalanced process—Self-directed teams will naturally provide the solution by shifting resources to the process steps that take longer. Process step 1 will now have the person from process step 2 helping the longer process step. They will cross-train each other, or through their problem-solving efforts they can resolve issues like long setup times.
2. Running out of raw materials—Self-directed teams are more likely to setup an efficient KanBan system for supplies and also more likely to be involved with supplier improvements.
3. Waiting for information—If this is an issue, self-directed teams will begin measuring it and visually managing it so that the leaders will see the impact during their daily Gemba walk.
4. Quality issues—Self-directed teams take ownership of quality assurance. They will have regular meetings with their internal suppliers and internal customers. They will likely have at least one team member skilled in Six Sigma, the best methodology for resolving quality issues.
5. Equipment maintenance—Self-directed teams are empowered through the autonomous maintenance program and can sense maintenance symptoms much faster than any other department, including the maintenance department!

The main benefit of a self-directed team organization is that they take on the process ownership and they realize the waiting time issues every day; it is not part of the "hidden factory" to them. They, more than anyone, know the frustration of waiting. By providing them empowerment and the required skills, self-directed teams have the best opportunity to address this waste.

SUPERVISOR TRANSITION TO TEAM FACILITATOR

After the decision is made to organize into self-directed teams, the process supervisor will also be required to go through a transition. This can be a sensitive change, as the traditional supervisor has always been told to use command and control methods to ensure the expected outcome of the production system. Organizing in self-directed teams is *not* an opportunity to remove this position. We are building an organization of support, development, and knowledge sharing, not an organization of minimum people requirements. Your opportunities for future cost savings do not come from a reduction in human resources (your most valuable asset); they will come from standard work improvements, uncovering process key points, process variation reduction, and employee motivation to continuously improve. The supervisor is very much needed in this new organization, but in a very different role.

I will now start calling the supervisor a *team facilitator* to avoid any confusion. The team facilitator will now also require support and reassurance in the new role. The team facilitator will require new skills such as follows:

1. TWI senior trainer to support the TWI trainers within the self-directed teams
2. Black or Green Belt to support and mentor the self-directed team's problem solving
3. Effective team meeting skills to ensure self-directed teams are conducting effective meetings
4. Team facilitation skills to support team development and effective meetings
5. Conflict resolution within teams
6. Of course they will already have the skills of work order management and ensuring the day-to-day production requirements, which they will continue to use but more in a mentoring role

Supervisor Transition Implementation Steps

1. Review the current supervisor job description. What must be changed to reflect the new tasks and skills? One task will be bringing the teams through the empowerment steps and teaching the skills required by the self-directed teams. The team facilitator will be supporting more than one self-directed team.

2. The new job description is developed by the Management Steering Team. Once completed, the communication of the new self-directed teams and their new role is communicated to them. The timing is critical, because if they hear rumors about the self-directed team organization, it will be more difficult to convince them of their new and valuable role.

3. The training program for the supervisor transition outlines the skill subjects, who will provide the training, and time lines. This is also given to the new team facilitators. It will be the responsibility of the Management Steering Team to manage this training to prevent any delays.

4. The Management Steering Team then prepares and communicates the new organization to all employees. The benefits and value of highly developed people are the main two messages.

5. Finally the Management Steering Team members must be prepared to provide mentoring for the skills required for this new position of team facilitator, so please become experts if you are not already. Self-learning is an attribute of a leader.

Self-Directed Teams: A Culture of Engagement

A 2011 Gallup Poll (Blacksmith and Harter, 2011) of American workers found 71% "not engaged" or "actively disengaged" in their work. If only 30% of your workers are involved, you must change the system. To what? Self-directed teams will provide a significant increase in engagement. Why? Because people build individual value and security by being a member of a self-directed team. They feel comfortable contributing. They have a sense of belonging because they feel worthy. Brené Brown (2010) is a research professor at the University of Houston and studies human connection—our ability to belong and empathize. She found the key to this was having a sense of being worthy. If self-directed teams are developed correctly where they strive for and obtain new skills, where they are taught and supported, and where they see that their contributions are inputs toward meeting the organization's single-focus strategy, they will soon obtain this sense of being worthy.

Self-directed teams have been around as far back as the 1960s; however, they have not really gained in popularity with organizations. This can only be attributed to the fact that they are very different from the traditional organization and culture. Traditional organizations were structured for compliance, but if you want engagement, self-directed is better.

I once was listening to an NPR radio program (2012) where they interviewed an engineer who worked at the Jet Propulsion Laboratory and he was a rover driver for Curiosity, the robotic vehicle used by NASA to investigate Mars. What was interesting was this engineer was average in school and was not a scientist. When asked how he got the rover driver job, he recognized the self-directed team he worked with at the Jet Propulsion Laboratory. He said they created an environment of curiosity that he thrived on. Within the team, if you showed an interest and were willing to learn, you were supported and allowed to gain the necessary skills if you were able to go through the team's "rite of passage." This was a great example of why self-directed teams are beneficial to an organization; they are able to bring out superstars who were previously sleeping.

Successful implementation of self-directed teams has many benefits:

- Individual value and security
- Productivity and quality improvements
- Higher motivation
- Environment of creativity
- Flexibility
- Increased employee involvement (Because of this, no need for a suggestion program; they develop and implement their own ideas.)

The task of the team is to manage, control, and improve their part of the process that can be defined by a manufacturing cell or an administration process within the office. They are given targets within these categories and are held accountable for the results.

The team must have a structure that is used mainly to conduct effective meetings. They are given one hour per week of nonproduction time for improvement. You will find that as the team matures, it may need to be empowered to have further improvement meetings during the week;

however, the team is always conscious of production responsibilities. The key team roles are as follows:

- Team facilitator (This is typically the supervisor and can manage three to five teams.)
- Team leader
- Team notetaker
- Team timekeeper
- Team maintenance
- Team TWI-JI Trainer

To provide further structure, and from the beginning, the team develops norms, which are 100% agreed upon by all team members and are used to ensure proper behavior and provide processes like submitting ideas, coming to a decision, individual behavioral correction, and the like.

Management must change first. They must understand and acquire skills on supporting and facilitating self-directed teams. One of the key changes for managers is to be able to accept that self-directed teams will make mistakes. If they are not able to accept this and act with appropriate coaching when a mistake occurs, there will be no building of trust. The first step in achieving engagement is achieving trust. Supervisors will transition to become the needed team facilitators, and they must acquire these abilities, teach a problem-solving process, teach Lean Sigma skills, conduct effective meetings, and teach teams how to resolve conflicts. As a team facilitator their main objective is to get the team to autonomy with all of the skills I just mentioned. Once this is accomplished, their new role is to be an expert in supporting self-directed teams in their improvement efforts. One of their expert skills will be to become a senior trainer for the TWI programs.

The Management Steering Team is responsible for developing the levels of empowerment; in my case we had five. Once developed, they use the Gemba STP methods (support, teach, promote) to help the teams go through these levels of empowerment. Management must hone their skills needed for STP, as they are key in developing this new culture. Examples of empowerment levels are managing their workloads, planning the work, Standard Work improvements, problem solving, completing their own appraisals, managing overtime, managing an operating report (costs), maintenance, developing their skills, and others. They are unique to each organization, but they must be developed because this is where individual value will come from, which of course also significantly contributes to beneficial engagement.

TEAM MEMBER FUNDAMENTAL TRAINING

Here I will use the word *teams* as an acronym to provide the fundamentals each team and team member must work toward achieving: Trust, Evolution, Accountability, Measurability, and Sustainability. You can break these out into a PowerPoint presentation to provide training for your self-directed teams. There are of course many variations of training programs on how to work within a team. In the 21 years of my career with self-directed teams, I have been trained in a few of these variations. Some of the teachings I have found invaluable and have practiced to this day. However, I found some fundamentals either not mentioned or not emphasized. Based on business results and from what I have experienced, here are the foundations of successful teams to be included in training:

Trust—If the team is going to have a single focus on what to develop to be successful, it would definitely be trust. Each member of the team must be able to trust the other team members—each and every one. They must be able to trust them to accomplish their day-to-day tasks, to adhere to the team norms, to share knowledge and key points, to support each other when mistakes are made, to effectively communicate, and to complete assigned tasks. With team members confident and reliant on each other's abilities, the team will achieve the expected extraordinary business results of teamwork. Without individual trust, the team will not function as a team and results will be pedestrian.

Evolution—The team must have both a team and a team member development plan toward achieving high performance. The organization going to a self-directed team structure must realize that team members will need higher level skills to make empowered decisions. These skills are to be aligned with the strategy and outlined within the team's career path. The team can then be empowered to develop a plan and path for each and every team member. This will provide a fast-paced team evolution toward high performance and achieving business results.

Accountability—All team members must be responsible for achieving assigned business targets for their process scope. Targets are aligned with the strategy, and each team member is clear on how they contribute. To be accountable they will need to be empowered (to establish ownership) with the authority to request information, engineering support, maintenance, supplier support—all of the resources needed to be successful. Team members must also understand that they must be accountable to their internal and external customers and to each other. A self-directed team structure provides customer interdependency throughout the value chain. Each team must understand their accountability for meeting the requirements of their internal customer and thus the external customer. Measurable goals must be met in all of the business requirements: Quality, Cost, and Delivery (QCD). Finally they must be accountable to each other by completing assigned action plans on time, supporting team decisions, and developing themselves.

Measurability—The team will have business targets assigned by the Management Steering Team. This alone demonstrates to the teams that they are clearly a key part of the organization. Meeting the targets will clearly demonstrate the weight of their contribution toward the success of the business. Daily and strategy measures are visually managed by the team. Being visually managed facilitates support from the leaders of the organization. Measuring and developing process inputs are key for process and continuous improvement at this level. Measuring only outputs provides stress and continued firefighting.

Sustainability—Along the path of team development, all accomplishments and business results must be sustained. Without sustainment, successful efforts become extremely wasteful and demoralizing. Achieved business results from team improvement projects are visually managed for a minimum of three months after the effective solutions have been standardized and put in place. Team skill development is sustained through regular team reviews of the each member's ability to follow the already established standard work. These reviews are also visually managed within the TWI matrix. All of these sustainment efforts are also supported by the tiered management Gemba walk. The complete management team must be clear that this newly formed organization requires their support, teaching, and promotion.

These are proven team member foundations not only from my experience but also from many other organizations that have successfully implemented a self-directed team organization.

TWI STRUCTURE

The Training Within Industry (TWI) programs—Job Instruction, Job Methods, Job Relations, and Job Safety—are very powerful and are easily adapted by self-directed teams. They are designed for practical use at this level of the organization and they bring results. What I am going to talk about here is the TWI-JI (Training Within Industry—Job Instruction). All of the programs were developed during the Second World War within the United States. The reason this program was needed and developed is that most of the men had gone to war, and women and unskilled labor had to fill in on the manufacturing floor to keep supplies going to the war front. Now most of these women had never worked before, so imagine the challenge in designing an effective training program where unskilled people had to learn job skills very quickly. The Job Instruction process accomplished this almost impossible task. So why didn't it stay around? Well, it did, but it went to Japan, as did some of the other process improvement techniques of that era. It is an integral part of the Toyota Production System (TPS), but it was obscure as they did not apply a label to it—they just did it. Toyota valued it as a program for getting standard work into place and as a method for discovering key points about quality, safety, and efficiency. The method of delivery by the certified TWI-JI Trainer is in fact executed in standard work. It is a four-step process (Figure 3.1), and TWI trainers are taught they must deliver according to this four-step method. Each trainer receives one of these "How to Instruct" cards. This standard delivery is to ensure that first the learner learns and second the training reduces process variation. The learner learns because the job breakdown performed again by the TWI-JI Trainer has certain effective rules, such as, the important steps delivered must be less than 10 (how many people can remember more than 10 steps at a time?); each important step may have key points; and if there is a key point, then there are reasons for the key points. Following are the criteria for determining key points:

1. Make or break the job (like a quality issue or creating a machine problem)
2. Could injure the worker (safety prevention)
3. Make the job easier to do (efficiency of performing the step)

The above criteria is what makes the TWI-JI program equally an improvement program. When workers start to realize the power of discovering these key points, they start to achieve a language and another method for

4 STEPS FOR JOB INSTRUCTION – How to Instruct

Step 1 – PREPARE THE WORKER

- Put the person at ease
- State the job
- Find out what the person already knows
- Get the person interested in learning the job
- Place the person in the correct position

Step 2 – PRESENT THE OPERATION

- Tell, show, and illustrate **one Important Step** at a time
- Do it again stressing **Key Points**
- Do it again stating reasons for **Key Points**

Instruct clearly, completely, and patiently, but don't give them more information than they can master at one time.

Step 3 – TRY-OUT PERFORMANCE

- Have the person do the job — correct errors
- Have the person explain each **Important Step** to you as they do the job again
- Have the person explain each **Key Point** to you as they do the job again
- Have the person explain **reasons for Key Points** to you as they do the job again

*Make sure the person understands. Continue until **you** know **they** know.*

Step 4 – FOLLOW UP

- Put the person on their own
- Designate to whom the person goes for help
- Check on the person frequently
- Encourage questions
- Taper off extra coaching and close follow up

IF THE WORKER HASN'T LEARNED, THE INSTRUCTOR HASN'T TAUGHT

FIGURE 3.1

TWI four-step process. (From TWI 4-Step Process, *4 Steps for Job Instruction*, pg. 4. Retrieved from http://twi-institute.com/. Reprinted with permission.)

improvement. I once benchmarked with a company that had implemented TWI-JI about two years earlier. During the plant tour the explanation of their program was proudly conducted by the workers. They allowed us to sit in on a discussion about a recent customer complaint, again conducted by all workers from the value stream. After the explanation of the root cause, one person asked, "Do we have a key point in our job breakdown that addresses the issue?" They then reviewed the job breakdown at the point of the process where the issue occurred, and sure enough, there was no key point to prevent the issue from happening. So they changed the document to include the key point and the reason for the key point. All workers in that part of the process were immediately trained only on the new key point and the reason for the key point—no need to be retrained on the complete job breakdown. The workers' identification of key points is a very powerful and effective part of the program because:

1. The workers within the process own the process.
2. The breakdowns are developed by the workers, which makes it their language.
3. The breakdowns are one page and not a book forced on them by an industrial engineer.

They let us know that the TWI-JI program was the only problem-solving method they used. In fact, once the workers take ownership of this program and start using it as an improvement program, they are following an improvement cycle to continuously improve their standard work instructions (see Figure 3.2).

What I have realized is that this program needs an organizational structure. Some organizations use a trainer pool. I have seen two problems with this: First, it usually generates a backlog of training requests, and second, but more important, the trainers do not feel ownership for the part of the process they are training in. The structure I support (Figure 3.3) is having a complete support system for the TWI-JI Trainers. The support system starts with the Management Steering Team using STP techniques during their daily Gemba walk. Of course it is key to have visual management of the TWI matrix in place. This is very important in providing a fast pace of the required training, because the matrix shows planned dates for the training; if a date is missed, then that box in the matrix is colored red and noticed on the daily Gemba walk. There can now be an immediate decision as to a new date, and it is colored green again. In a traditional system the trainer misses the date and

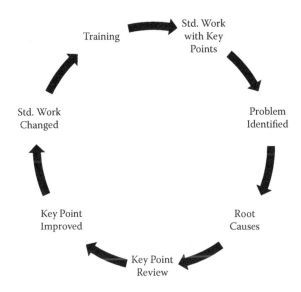

FIGURE 3.2
Standard work development cycle through key points. (From Robert Baird.)

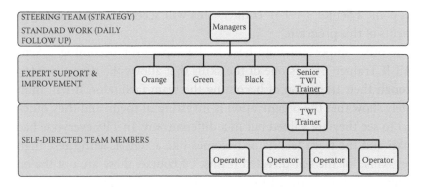

FIGURE 3.3
TWI organization structure. (From Robert Baird.)

no one knows or people are too busy with other issues to determine a new training date. This alone increases the cycle time of training. The next level of support is the team facilitator certified as a Senior TWI Trainer. A Senior TWI Trainer receives a more intensive five-day training program geared to provide an expert level of TWI-JI. It also allows them to certify TWI-JI Trainers with the 10-hour course. The TWI-JI Trainer then has an immediate support system while they are honing their TWI-JI skills. The final level of support is a peer-to-peer support because the TWI-JI Trainer is part of a self-directed team and is responsible for training in all aspects of the process and machines using TWI-JI. It is peer-to-peer because all of the self-directed team members have been trained with the 10-hour certification course.

The beauty of training with the TWI-JI training method is that you do not need to know the tasks you are training on; in fact, it is better to not know because you are required to use only the words on the job breakdown. Doing this prevents process variation because each and every operator gets trained *exactly* the same way. Think of delivering a PowerPoint presentation: You do not use the same words for each group you present to. You miss key points with one group but not another. So did all groups get the same message or training?

Why is the self-directed team considered support for the TWI-JI Trainers? As I noted, they become trained in the four-step process and especially the key points, which become common language within various team meetings. With this team ability, they start improving their standard work and the job breakdowns. Now it is not only the responsibility of the TWI-JI Trainer, but one team member is identified as the trainer expert.

I also support having each and every team member receive the 10-hour TWI-JI Trainer course. This can be provided by the supervisor, who is at this point a Senior TWI-JI Trainer. This will accelerate the learning and benefits of this program.

The other important note here is the supervisor becoming a Senior TWI-JI Trainer. This is one of the skills that allows the supervisor to go through their transition of becoming the team facilitator. It will immediately show that the organization is investing in them, and they quickly start to see they are valued but in a different way. In TPS everyone had a Sensei and the team facilitator becomes like a Sensei, with these higher level skills, for the self-directed teams. Of course, these are not the only higher level skills planned for the transition of the supervisor. We have already discussed more about these supervisor transition skills in the Self-Directed Team Organization section, earlier in this chapter.

TWI STRUCTURE IMPLEMENTATION PLAN

With an organization that includes self-directed teams or even a system of Kaizen event teams, the TWI structure is fundamental in achieving fast-paced improvements and standard work. The teams easily understand the TWI programs and provide them with another program for improvement, whether it is process or people improvement. However, a successful program needs a support structure. This is how to put it into place:

1. Start with the Management Steering Team determining a business need or required result for a certain section or the complete value stream. This will be the pilot area. Determine the visual management to use during this pilot—which metric best represents the results you are looking for.
2. Select the responsible self-directed team and communicate the objective. Obtain their buy-in with "what's in it for me" and the business result required. They will now know how they will contribute.
3. Train someone from the Lean Sigma or quality department to the level of Senior TWI-JI Trainer. This must be the department manager. This trainer is required to provide STP to achieve full benefits from the TWI programs. In the future the senior trainer will be the team facilitator and will provide STP for the teams they are responsible for.
4. The senior trainer now trains all of the selected self-directed team members as TWI-JI Trainers. All of the members trained might not perform the actual training, but it is very helpful for them to understand the program and how it benefits their team. This training will also include the responsible team facilitator, who will be required to complete this 10-hour course to become a certified Senior TWI-JI Trainer. They also will be part of this pilot effort.
5. Start setting up the TWI structure by assigning who on the self-directed team will be the actual person or people performing the job breakdowns and subsequent training.
6. Provide a job breakdown matrix of the pilot operation. How many job breakdowns will be required, by whom, and by what date?
7. Measure the current situation for the business metric assigned by the Management Steering Team, and visually manage at the teams metric board.
8. Determine a validation process for the job breakdowns generated: Job breakdown by TWI Trainer > Reviewed and adjusted by the Senior TWI-JI Trainer with the TWI Trainer > If quality, safety, or engineering tasks, they are reviewed by appropriate department > Begin training of first worker > Adjust from what was learned at this training.
9. Develop a training matrix with job breakdowns to be trained, who will be trained, which trainer, and who and when the observations will be conducted of the people trained. Visually manage this matrix to ensure the dates are completed according to the plan; if not, then

immediately change to the next date. The matrix and date changes are managed by the supervisor/team facilitator and TWI Trainer.

10. Inform the Management Steering Team the training is about to start so they can start their STP of the pilot program during the daily management Gemba walks.

11. TWI-JI Trainers start the required job breakdowns. Many people make the mistake of completing all of the required job breakdowns for the targeted operation, but an important part of the validation process is training the first worker and making adjustments to the breakdown before going ahead with training everyone. This process is continuous: Job breakdown > Validation with senior trainer and appropriate department (quality, safety, engineering, etc.) > Validation by training the first worker > Did the learner learn?

12. During the job breakdown process the key points of each task are fleshed out. This is why you will be able to improve the business metric assigned by the Management Steering Team, because these key points—which make or break the job (prevent quality or efficiency issues), make it easier to do the job (improve cycle time), and could injure the worker (prevent a safety issue)—will now be known by everyone. It is the new standard work; but don't stop there! Keep reviewing these key points with the self-directed team during the pilot. Try to determine other key points that will have an effect on the result you are looking for. If you find one, update the job breakdown and retrain the workers on this important step and key point only; no need to retrain on the complete job breakdown.

13. With the results from the pilot achieved, the self-directed team presents the success to the Management Steering Team. These results are then celebrated and communicated by the self-directed team to the other teams.

14. All supervisors/team facilitators are trained as Senior TWI-JI Trainers.

15. With these new skills the Senior TWI-JI Trainers (supervisors/team facilitators) train and certify the self-directed teams they are responsible for with the 10-hour TWI-JI Trainer course. The supervisors/ team facilitators are now responsible for supporting and developing not only the program but also the development of the key points and progressing standard work to the highest level possible.

16. With all of the required positions trained, the Management Steering Team assigns a team to determine an organizationwide job breakdown matrix (same as in Step 6 but for the complete organization).

17. With the job breakdown matrix determined, each self-directed team develops the job breakdowns for their part of the process and starts training according to the training matrix. This training matrix is now visually managed at each self-directed team's three-sided metric board, and progress is supported by the daily management Gemba walk.

18. As more and more results are realized and standard work is improving, then it is time to look at implementation of the other TWI programs. My recommendation is to start TWI-JR or TWI-JM next. You now have the required support structure in place to provide fast-paced progress so the benefits from implementing the next TWI programs will come much faster.

HARADA METHOD

The Harada Method is a relatively new method, since about 2002, which is very much aligned with what I talk about throughout this book: development of people and teamwork. It also is a method that promotes improvement *every day*. It is very successful because it is about the betterment and development of the individual. Of course, managers play a key role in the plan and in the support of these individuals reaching their goal. They have to be able to teach people how to be successful.

The Harada Method focuses on developing people to their maximum creative capacity and team development. This results in quality, cost reduction, innovation, and profit. Developed by Takashi Harada from Japan, who was a junior high school teacher, this method brought the school's athletes from being the worst ranked in Osaka to the best, five years in a row. He is now a consultant and has trained 55,000 people in 280 companies in the Harada Method.

Principles of the Harada Method

- Self-reliance—People do not need much support in working toward their goal.
- Goal-oriented—People start by picking a goal for themselves, what they want to become an expert in. The goal is aligned with their aspirations and the company's success. Then when people come to work they will have a goal to continuously develop. They are clear on what

has to be achieved. They study and obtain skills that enable them to achieve their goal. They have a plan to achieve these needed skills. They are passionate about achieving their goal.

- Develop people to their fullest potential—The Harada Method states that everyone can be successful. The time line for development is developed by the individual. They also create the necessary measures and how to monitor.
- Everyone can be successful—No one is left out of this process. It is not only for the high-performing people within the organization. The objective is developing all people to their highest potential.

Harada Method Steps to Self-Reliance

- Goal
- Purpose—Why do you want that goal?
- Analysis—Review past successes and failures.
- Action—Then put a plan together, from how to avoid the failures, what were the weaknesses and what made the successes, how can they be repeated?
- Routines—What are the 10 daily routines to establish new habits that will make you successful?

I am not a certified Harada Method instructor currently, but I am planning to become one. To get started on this very powerful program, contact PCS Press at http://pcspress.businesscatalyst.com/?p=382.

JISHU HOZEN (AUTONOMOUS MAINTENANCE)

Autonomous maintenance performed by the self-directed shop floor teams is very effective in reducing process variation, improving process cycle time, and improving quality. Lower machine efficiency can have a significant effect on cycle time because of buffer WIP required at these operations.

The method I explain below is different from the standard Jishu hozen, from the standpoint that the self-directed team is focusing more on building skills to complete the small repairs. I have found that these small repairs contribute to 60% of the overall repairs of the machine. The downtime savings come from removing most of the steps of the complete maintenance cycle: machine stops > work order generated > wait for technician to arrive > repair > validate repair. The steps that are removed are wait for technician to arrive and validate repair. I have found the waiting for technician step is up to 85% of the total maintenance cycle. Once the self-directed teams are completing 60% of the repairs, the maintenance technicians can now focus their expertise on preventive maintenance. The self-directed teams are still responsible for some preventive tasks associated with the 5S program. The related tasks here are scheduled cleaning of the machine, but not just to improve the appearance. From a review of the maintenance database with the maintenance technicians, the cleaning can be much more effective by identifying where they had to clean during a repair. These areas can then be color coded and then put on the 5S schedule for frequency of cleaning. Every shift or daily is not an acceptable cleaning interval; we can get more scientific with machine counters and electronic display systems.

The self-directed team members actually adopt Jishu hozen very well, and it can also be one of the new skills to be achieved through their career path. To have it work, the maintenance department must be part of the implementation. This department can be more sensitive because they become responsible for the training of the self-directed teams on some of the repairs and tasks they used to do. This is the reason, before starting the actual review of what repairs and tasks the self-directed teams will now be responsible for, clear communication is needed on why the organization is going in this direction. There are two main messages that can be communicated:

1. Some operations have known unacceptable capability, resulting in unacceptable lead times for our customers.
2. We need to use the higher level skills of our maintenance staff for preventive tasks.

There can also be other messages tied to what the organization has selected for the single-focus strategy, but these two are very understandable and the second one reassures that the organization is not trying to find ways to reduce the maintenance staff.

How does it work? There are three basic steps to complete:

1. Review of past repairs
2. Identification of repairs and training of these tasks
3. Follow up on the effect

Review of Past Repairs

This is done to determine which repairs self-directed team members will be able to complete safely and effectively. They are sometimes repairs that the team members have taken the initiative, in the past, to complete on their own. So review with the team members to flesh out these types of repairs.

Identification of Repairs and Training of These Tasks

This review uses the maintenance database. A search filter is used to determine repair times of less than 15 minutes. Repair times less than this indicate a less complex repair. After the repairs have been identified, job breakdown documents are developed for each repair, using the TWI-JI training method. Finally, it is the responsibility of the maintenance person who determined the breakdown to deliver the training according to the visually managed TWI training matrix. Of course at this time the maintenance staff involved in the autonomous maintenance program have received the TWI-JI 10-hour training program delivered by a Senior TWI-JI Trainer.

Follow Up on the Effect

Metrics need to be generated like WIP levels, machine downtime related to maintenance, overall equipment effectiveness (OEE), and others. These are visually managed at the cells using the autonomous maintenance program. They are then reviewed during the managers' daily Gemba walk. The Management Steering Team also reviews the results at their weekly meeting.

Jishu Hozen Implementation Plan

Autonomous maintenance performed by the shop floor self-directed teams and the maintenance staff implementation and culture is the direct responsibility of the Management Steering Team. They will need to take the proper steps and then ensure that business results are sustained. They will also need to align the effort with the single-focus strategy to facilitate the communication to other stakeholders. Here are the implementation steps you can follow:

1. The first step is to have the Management Steering Team—
   a. Determine how it aligns with the single-focus strategy.
   b. Determine what metrics to use, their current levels, the targets, and how it positively impacts customers.
   c. Prepare a communication designed for all of the stakeholders that includes the information from the above points.
2. Management Steering Team communicates to all of the stakeholders.
   a. The nine-minute meeting rules can be used here.
3. Determine who requires training, who will provide the training, and content of training.
   a. Maintenance staff and the self-directed teams all need to be trained on TPM, especially the autonomous maintenance.
   b. If not done already, the maintenance staff will need to receive the TWI-JI 10-hour training by a senior trainer. This will provide them with certification to deliver effective and efficient training.
   c. The training should be delivered by someone within the management team.
   d. The content should be such that the training should not take longer than 20 minutes. It must also include the objectives and explain how it aligns with the single-focus strategy as determined by the Management Steering Team.
4. Form the implementation team.
   a. A3 is developed by the team and reviewed by the Management Steering Team for alignment.
   b. Project review time line is determined. The milestone of solutions implemented should be set at one month. With the support of the Management Steering Team and the daily Gemba walks, this milestone time target should easily be met.
5. Decide which repairs and/or tasks can be performed by the self-directed teams.
   a. Search the maintenance database for repair times less than 15 minutes. This is the actual repair time, not the complete cycle of machine stopped, work order generated, maintenance tech arrives, repair time, and work order closed.

    b. These repairs are then reviewed to determine safety and complexity. However, if the organization has been using TWI-JI for at least a year, it should not be too much of a concern, and most repairs should be accepted as candidates.

6. Generate the TWI job breakdowns for each of the repairs.

    a. Job breakdowns are completed by the maintenance staff.

    b. The job breakdown is recorded by a maintenance tech, and the person completing the repair is another maintenance tech. The maintenance tech observing and recording the important steps, key points, and reasons for the key points will deliver the training, and the maintenance tech being observed doing the repair must know the repair well to bring out the key points.

    c. The job breakdowns go through the validation process with the senior trainer and any required experts (safety, engineering, quality, etc.).

7. Start training the self-directed team members.

    a. The TWI training matrix is developed by the maintenance tech executing the training and the team facilitator for each shift. It is then visually managed at each of the cells who will receive the training.

    b. The first team member is trained to validate the job breakdown.

    c. With validation, all applicable team members are trained.

8. The self-directed team is responsible for manually recording the completed number of repairs they were trained on.

    a. If the maintenance database can flag these repairs, then it would be the responsibility of the self-directed teams to review and chart against targets of time to repair and number per units produced.

    b. The next step for the self-directed teams is to reduce the number of repairs per units produced. This step will probably take six months before starting.

9. Management daily Gemba walk to apply STP techniques—

    a. Review the visually managed metrics at each of the cells.

    b. Support, teach, and promote this program.

10. Metrics set to show the program is sustaining results—

    a. Management Steering Team to work with self-directed teams on which metrics are best visually managed to indicate that results are sustaining.

LEAN SIGMA STRUCTURE

To obtain results and, more importantly, sustain the results, there must be an organizational structure of support for key inputs to provide a continuous improvement culture. Too many organizations waste a considerable amount of investment in Lean Sigma training only to realize one-time results and then put the skills in the closet. There is really no question by management as to what happened to this investment, and everyone goes back to what they used to do. This can be a considerable one-time cost, and the loss for potential savings is even greater. In a benchmarking study done in 2006 by the Aberdeen Group, only 16% of companies with Lean Sigma initiatives are holding fast to the rigorous programs of Lean Sigma. This is because there is very little organizational structure to support and coach these new continuous improvement resources. Also, without the Management Steering Team to ensure that these people are consistently leading and responsible for the key improvement projects, there is very little importance or organizational value felt by these higher skilled people. Organizational structure starts with the Management Steering Team (see Figure 3.4). This team provides the individual value, which provides the motivation and passion to do well. The Master Black Belt is there to initially provide the training, but their main responsibility is to be a mentor

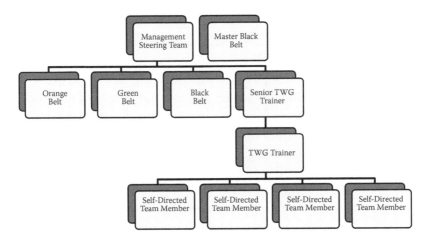

FIGURE 3.4
Lean Sigma organization structure. (From Robert Baird.)

for the people they have trained. I found low certification rates and continuance of practicing after training because the newly trained belt does not instantly become a practitioner after the classroom training. They require further support, coaching, and encouragement as they are honing their skills. Sure, some people become practitioners on their own, but only a very low percentage. It has nothing to do with their intellectual ability to comprehend the techniques; it is more to do with the methodology of application, and this is mainly where the Master Black Belt is required.

New Product Introduction

Passing a new product from R&D to manufacturing with low manufacturability always results in significantly higher costs for any organization. This is because the defects continue to fall out for the life of the product. Defects that are not caught by the manufacturing quality system go to customers, and the cost significantly goes up again and revenue is lost. In the Toyota Production System, new products must be proved before they are released into production. This means they have to be tested to prove they meet an acceptable capability. In late 2009 Toyota's most popular model, the Toyota Camry, had to have a massive recall related to unexpected acceleration (CBS News, 2010). The top executives of Toyota ended up having to testify to the U.S. Congress. What was interesting was that their answer was not to increase inspection or quality control at the manufacturing step but to review how they can improve quality at the concept stage. They understand this is where quality is built in and saves considerable costs and new product introduction cycle time. Later Toyota admitted they were victim to a focus on growth and had strayed away from the principles that had made them very successful.

Designing new products and machines to build quality in must be the single-focus strategy. Too many organizations have the single focus of cost reduction, which is a critical mistake. With a cost reduction single focus, proven specifications are being challenged: Can we use less material, can we use a cheaper material, and can we accept more variation and wider upper and lower specifications? Do not get me wrong. These questions should be asked, but they need to be scientifically proven.

Interdisciplinary teams or more specific interdisciplinary thinking is the new process of developing an acceptable product or machine. Why? There are new challenges, current problem solving is becoming obsolete, it is the best path to innovation, and it will save time and resources (cost). It must be started at the beginning of the process. I worked with an organization that tried to start the interdisciplinary team at the time the new product was already at the stage of developing. The team members never did become a team because they already had their assignments and it was difficult to bring their disciplines together. Bringing together different disciplines brings different perspectives. Communication and collaboration skills become very important. In a team environment people will be thinking, "Why don't they see things the way I do"; "I am afraid of the others' reaction because I am not familiar with the other disciplines"; and "Listen to my idea—it is better." On any team, communication feedback and understanding of the roles are key. With interdisciplinary teams it is critical to provide a team facilitator to ensure these important elements happen and to provoke their creativity. Another part of the team is the customer, who is required for instant feedback during the prototype period, just before going to manufacturing, and at the time of customer reception. Just by their involvement alone, look for a 30% reduction in cycle time and 25% improvement in quality.

Some team structure must be established: how each member contributes, the roles within the team. This must be established as not all disciplines will have the weight. People must know what they have to do, how, with whom, and how to engage. The individual team members must see how their contribution fits into the outcome.

DYNAMIC PROCESS IMPROVEMENT (DPI)

Creating a culture of fast-paced improvement requires a process and organization designed to easily dig out the root causes of key process input variation. Dynamic Process Improvement, or DPI as I have labeled it, requires three key elements:

1. Self-directed teams
2. Dynamic problem-solving process
3. Visual management

Self-Directed Teams

Here is a bold statement, but true: Most organizations are capable of process variation reduction and waste removal. The problem is the resources required to accomplish these objectives. It is impossible to be cost-effective in resolving the root causes of variation and waste with the typical number of process engineers and Lean Sigma people. Achieving self-directed teams provides a cost-effective solution and fast pace toward business results. Developing an organization of self-directed teams at each key process step provides ownership with the responsibility of continuous improvement. These teams are made up of the people already running the process, and because they are there 100% of the time, they know the process better than anyone within the organization. They know their suppliers and customers, and are measuring performance aligned and required for strategy deployment. When supported with skills, information, and authorization to request needed resources, self-directed teams will help create a fast-paced and dynamic process improvement culture.

Dynamic Problem-Solving Process

The common and understood problem-solving processes today are the Six Sigma DMAIC, the Lean PDCA, and the steps of completing an A3 report. They all follow the basic steps of problem identification, root cause analysis, solution development, and confirmation that the problem has been reduced or eradicated. Following DPI basically requires following these steps but in real time, every day, and every shift. One of the main problems of DMAIC, PDCA, and A3 is that they use data collection and root cause analysis techniques sometime after the problem has occurred, when the root causes are no longer fresh. Tools like SPC, histograms, Pareto, design of experiments (DOE), and root cause analysis are interpreted too late, and even the best practitioners can treat the symptom and not the root cause.

With a dynamic problem-solving process, the self-directed teams are trained to correctly define the problem, provide a current measure and target, identify the key input variables utilizing an IPO (Input, Process, Output) diagram, provide solution ideas for the root causes of these key input variables, create standard work, and sustain the results. The table shown in Figure 3.5 is completed and visually managed on one side of the self-directed team's three-sided metric board. The key part of the process and why it is dynamic, in real time, is that when the key input variables are identified, they become the first level of the root cause analysis. As the self-directed team members are working and issues start arising, the team members are writing down their observations immediately. They do not have to try to remember what happened at the end of the shift or even at the end of a pitch. They also use the hourly pitch chart, and each time they miss their hourly target (or in this case batch cycle time), they ask why for the appropriate key input variable. The

Problem Description: The Cycle Time of assembly 212 is causing late deliveries for a key customer	Input WIP	Input Machine	Input People	Solution Ideas
Current Measure: (chart: values 0–5 over Day 1, Day 2, Day 3, Day 4, Day 5, Day 6, Day 7)	Observation Upstream operation pushed two trolleys more than required	Observation Machine stops every 30 minutes	Observation Need cross training to cover for breaks	Kanban system, pull system
	Observation Run out of WIP	Observation Too many defects	Observation Shift change too long, 35 minutes	Use Spider to get next order
	Observation Waiting for Supervisor	Observation	Observation	Training required for repairs
	Observation	Observation	Observation	
	Observation	Observation	Observation	

Input Process Output

	Assembly 212	Cycle Time
WIP		
Machine		
People		

FIGURE 3.5
Team DPI table. (From Robert Baird.)

observation is once again immediately recorded. The team might even hold a Kaizen if they feel they have uncovered a significant cause and perform a small experiment with a potential solution. Once the table is completed or observations have seemed to slow, the team will then perform a root cause analysis with the key observations noted and determine the solutions. Some of the solutions usually come from the Solution Ideas column.

The metrics and targets for the metrics come from the Management Steering Team and are aligned with the single-focus strategy. Because the self-directed teams, by this time, have ownership of their process, they adopt the responsibility of meeting these targets, which contributes to the organization's already fast-paced problem solving.

DPI also encompasses the complete value stream because the self-directed teams own each part of the value stream or they own the complete value stream, depending on the number of steps in the value stream. With this type of value stream and continuous improvement coverage, problems related to the single-focus strategy are realized quickly.

This process, of course, is supported by STP during the manager's daily Gemba walk. If the plant layout is in streets and avenues and if the visual management is set up correctly at each of the self-directed team stations, then the Gemba walk can effectively ensure project progress within a 15 minute walk.

I would also like to mention something about recording of issues and downtime. Please do not use codes for downtime. I know it makes it easy for the people to record with bar-coding equipment, but two significant problems arise: First, if a new problem becomes apparent, the people having to record it either try to fit it into a category or, and this is the second problem, they enter it as *other*. Who knows the root cause of *other*? I have seen where the supplier started having a new material issue and the operator just kept entering the problem as *other*. This was before DPI was implemented, and with the old problem-solving process we calculated that our yield loss went up 1% for at least one month! So the best method is no codes and the team members manually entering what they observe.

VISUAL MANAGEMENT

Visual management is absolutely critical to a fast-paced organization. It must be in place before you can have an effective Gemba walk by the managers. Once you start the Gemba walk and you notice some of the managers are skipping going to the Gemba some days, you must ask if you have the correct visual management in place. Managers will not complete tasks that do not provide them with any value, so if they cannot easily see the project progress, progress toward the single focus, they will not support the Gemba walk, especially not daily.

At each of the key process steps owned by a self-directed team, there must be a standard three-sided board. The first side is used to dig out the problems and root causes by recording hourly process outputs. A target is set for these pitches and, if not met, then the reason is recorded, a Pareto is established, and a matrix is provided for the time of day the pitches are not met. Many organizations start with the obvious measurement of production volume or in services transactions, and this will provide a method for digging out the process problems. When I would promote this method globally, managers liked it, but only because the teams were measuring production output—and this is *not* the objective. It is a method to dig out the problems. The managers would always discuss why the team did not meet the shift production goal instead of reviewing the problems they were identifying. This misunderstanding will not help moving, especially at a fast pace, toward the single-focus strategy. In fact it will slow progress because focusing only on output metrics will add stress to the teams. When the managers are discussing the Pareto of problems by using STP, they are focusing on the improvement of the process and the related single-focus strategy. This is what contributes to a fast pace. The measurement of production pitches are not the only pitches that can be measured. For example, if the single focus is related to quality, yield loss per hour with a target could be measured, or if cycle time (contribution to lead time) is the single focus, then cycle time with a target per hour could be measured.

The second side of the standard three-sided board is used to record the process of Dynamic Problem Solving and the third side of this standard board is for sustaining the results by measuring and monitoring the progress of the resulting standard work (TWI recommended). The data chart used to measure the results of the Dynamic Problem Solving is continued for three months to see if the results are sustained. It is critical that this board is manually updated by the self-directed teams. Some people have come to the conclusion that this manual recording takes time away from the teams running their process. This time actually adds value as it results in fast-paced process improvement because of the culture developed from ownership. These measures are then reviewed, using STP techniques, by the appropriate managers during their daily Gemba walk.

Effective visual management does take some developing, but the three-sided board standard I outlined above will provide the value you are looking for. During the daily Gemba walk you should always be asking, Are the visuals telling us what we need to know? The daily Gemba walk will also be needed to quickly get it into place and become a culture of accurately keeping the metrics up to date.

STOP WORRYING ABOUT THE OUTPUTS

All organizations work hard to improve process outputs of Quality, Cost, and Delivery. All of their information systems can provide a wealth of data on the trends and distributions of these outputs. Their levels are well known among the interested managers. MBOs (Management by Objectives) are set every year to achieve improvements in these key outputs of the business. They are cascaded down to the lowest levels, which have little opportunity as individuals to influence these outputs. During my visits to various operations around the globe, I like asking managers how they think these lowest levels of the organization can provide improvements: Will they work harder, will they receive more empowerment, do they have the skills, or are they self-motivated enough to organize in order to provide the needed improvements? The answer is almost always the same: They have never really thought about it.

If organized as self-directed teams, the answer is obvious: They are empowered, they are skilled, and they do put more into their work when considering process improvements. However, if this level of the organization is focused only on the outputs, it will take much longer for the improvements to come. They must understand and know very well the inputs that influence their process outputs. This is where they must focus their improvement efforts and data collection through tools like SPC. Some managers focus on the outputs and have all kinds of nice-looking displays that automatically update process outputs. These might gain the interest of the people working in the process, but what do they accomplish? When visiting operations with such displays I like to ask managers if the people working in the process really know what these fancy displays are telling them. Their answer is of course they know and it helps them in their day-to-day work. So then I ask two or three people working in the process to provide an explanation of the charts and data tables being displayed. I have never received an answer where they can explain these data adequately to be able to influence the trend. I then let them know their answer is expected as we in management are doing a very poor job in using process data. A wiser approach is to understand the inputs influencing these outputs and to set process input goals instead of the output targets.

So stop worrying about the outputs—they really only create stress. Focus on and improve the inputs and watch your outputs improve.

REFERENCES

Aberdeen Group. 2006. *The Lean Six Sigma Benchmark Report.* (September). http://aberdeen.com/Aberdeen-Library/3490/RA_SixSigma_3490.aspx

Blacksmith, Nikki, and Jim Harter. 2011. Majority of American Workers Not Engaged in Their Jobs. Gallup Wellbeing Poll, October 28. http://www.gallup.com/poll/150383/majority-american-workers-not-engaged-jobs.aspx

Brown, Brené. *The Power of Vulnerability.* (2010, December) TED website. http://www.ted.com/talks/brene_brown_on_vulnerability.html

CBS News. 2010. Toyota Massive Recall Snowballs. (January 28) http://www.cbsnews.com/2100-500395_162-6149712.html

Harada Method course. PCS Press website. http://www.pcspress.com/?p=382

May, Michelle. n.d. *Self Managed Teams: The Organizational Form of the 21st Century.* TechCastArticleSeries.http://www.techcast.org/Upload/PDFs/060619084842TC%20Michelle.pdf

NPR radio. June 2012.

TWI 4-Step Process. *4 Steps for Job Instruction*, p. 4. Retrieved from http://twi-institute.com/.

Williams, Ron. 1995. Self-Directed Work Teams: A Competitive Advantage. *Quality Digest* (November 1). http://www.qualitydigest.com/magazine/1995/nov/article/self-directed-work-teams-competitive-advantage.html

4

Component 4: Fast Knowledge Sharing

Fast knowledge sharing will provide the organization with a wider distribution of knowledge. This wider distribution will facilitate pace in innovation, continuous improvement, and preventing issues from starting in the first place. It is really about becoming a learning organization and must be stated as one of the organization's values. It cannot be just one of the methods we use; leaders must include it and promote it as part of the organization's culture. Too many organizations are satisfied with providing sound and secure technology to store the knowledge. What they do not clearly understand is that the *people* are actually storing the expertise of the organization. This can be a catastrophic assumption, as the sound and secure technology might not fail but the unhappy experts can be enticed by the competition. Creating individual value by developing your people and keeping them challenged is how you keep organization knowledge and further develop expertise.

Each and every individual comes with and develops their own knowledge. The challenge is to learn how this development of knowledge is done and how the organization can effectively use it. We need to know what people know and then how to share.

Knowledge sharing is not just about technology; it is about people interacting. The organization must provide these interaction opportunities; it is part of the cost of improving knowledge sharing. If knowledge sharing is one of the organization's values, then there must be a budget for it. Knowledge sharing must also have key indicators in order to understand how it is done, the frequency, and the financial benefits to the organization. Knowledge sharing is about creating value.

Knowledge sharing will facilitate innovation, providing new products and services. With diverse knowledge, each individual increases their opportunity to provide ideas for innovation. It will increase their ability for collaboration, getting them out of a tribal culture.

Process knowledge sharing has many benefits, and your organization must build a learning organization. Leaders once again play a critical role in structuring the organization to learn.

This is an element of Lean Sigma that does not get much discussion but soon it will become (if it is not already) a significant competitive advantage. I have found underutilized knowledge to be one of the greatest wastes in any organization, and for many Lean practitioners it is the eighth waste:

1. Overproduction (faster than necessary pace)
2. Waiting
3. Transport
4. Inappropriate processing
5. Unnecessary inventory (excess inventory)
6. Unnecessary motion
7. Defects (correction of mistakes)
8. Underutilizing people's knowledge, talents, and skills

If you think of information as a required material for your product, knowledge as a waste might start to make sense. For example, you have one or two people in your organization who might be the go-to people when one of your high-technology pieces of equipment is down. So let's look at this in comparison to required process material that must be in place just in time and without error. In the first scenario, a process team member is running your high-technology machine but does not plan for the required volume of material to complete the day's schedule and therefore has to stop the machine to obtain more material. Waiting is the waste, and not having the required volume of material is the first *why* in the cause analysis. In the second scenario, a process team member is running the same high-technology machine and comes across a software problem, which subsequently shuts down the machine. Waiting is the waste, and not having the ability to diagnose and repair the machine is the first *why* in the cause analysis. In the first scenario the required material had to be transferred to the machine, and in the second scenario the required knowledge had to be transferred to repair the machine. Both scenarios had the same unnecessary effect: the machine was stopped because of something missing.

Please do not get me wrong here: There are many repair types that the process team member will not be able to do because of technical complexity. However, there are many more repairs and adjustments that a process team member is capable of completing if the training is provided (knowledge transfer). We came to this realization after collecting time data within the maintenance request process. The scope was from the time a maintenance request was started to the time the maintenance request was closed. Here were the intervals we tracked:

- Maintenance request opened
- Maintenance technician arrives at the machine
- Maintenance technician completes repair
- Maintenance technician closes the request

Can you guess which time interval took the longest? It was the time interval between the maintenance request being opened and when the maintenance technician arrives at the machine. The next interval, from the time the maintenance technician arrives at the machine to the time the maintenance technician completes the repair, was another interesting revelation. In more than 60% of the requests, this interval measured less than 15 minutes—an indication that these repairs were not very complex and could be completed by … yes, the process team member running the machine.

The solution to this problem was to share the maintenance technician's knowledge with the process team member. Using our Total Productive Maintenance (TPM) Team, we reviewed all repairs that took less than 15 minutes to complete and then determined if process team members could safely and effectively complete these tasks. The next step was to train the process team members in the completion of these repairs and then to make these repairs a team member skill as part of their career path.

The results were very impressive! Prior to the solution we were measuring the number of requests per million parts made, which was measured at 300 requests per million parts. After the solution had been in place for six months, the requests per million parts made came down to 12. We also measured the availability of our machines across a shift. This number averaged at 68% of a 7.2 hour shift before we put this solution into place. Again after six months, with this one solution, we increased this number to 81%.

Another part of the knowledge-sharing element is having, for every manufacturing cell, each process team member becoming an expert in every part of your process. For example, Team Member 1 might become the expert on machine A and Team Member 2 might become the expert in your statistical process control (SPC) program. These experts now become the technical resources for the team. I realize that these team members will have limitations in the depth of knowledge, but so do process engineers, who have to seek further information from other resources. The process team members would do the same. There are several advantages here:

- The process team members communicate in the same language.
- There is no waiting; the process teams now have a resource close by.
- Problem-solving abilities are enhanced and problem-solving time is shortened.
- Personal motivation and pride increase.
- Ownership of the process increases.

I described these examples to provide you with a different way of thinking of knowledge sharing and the benefits it brings to the organization.

If you have a fast-paced learning organization, four other programs can facilitate the benefits of knowledge sharing and create the fast-paced competitive advantages of the new age:

- Worldwide accessible process database
- Process standardization
- Social networking
- Succession planning

WORLDWIDE ACCESSIBLE PROCESS DATABASE

A project database is very powerful for sharing knowledge, which saves time and valuable resources. This statement holds true if there is a well-designed process to share the knowledge. Of course, any process must be measured, and the IT department is responsible for providing measures of how this uploaded knowledge is being used. Many organizations use SharePoint software to store knowledge but ask the people, who are sometimes technical experts, where their uploaded files are located—and even the technical experts have trouble finding it. The knowledge-sharing process must include a rating system to identify knowledge that can be shared, how it was developed, who the experts are, and how this critical knowledge will be communicated. People around the world must also be able to effortlessly search for similar projects that were completed successfully by another location. In some cases a successful project by another location can be "copied and pasted," significantly reducing the resource costs of starting a project from scratch.

A few software products are on the market to provide a depository of knowledge, like PowerSteering, Quality Companion (add-on of Minitab), TrackVia, and others, which can be used for global access of your successful projects. These packages must be flexible to meet your organization's needs, provide fast access from any location around the world, be secure, and be user friendly. The software package should also be able to provide project statistics like project cycle time, current project status, project cost/investment for ROI, savings (hard and soft), savings by your unique fields, savings by project leaders/Six Sigma belts, proposed savings, number of projects by your unique fields, and others. From these statistics you must build standard reports for the different levels, including the shop floor and executives. The standard is determined by talking to these levels directly to understand what is of value to them. From there the reports can be set up for automatic distribution; frequency is determined by each of these levels. I also recommend having someone in each location to present these statistics during key communication meetings, again at all levels.

To get started, each of your locations should have an assigned champion to—

1. Learn the use of the software
2. Be the administrator
3. Provide the training to all users

To set up the software to a standard process that will be used by all locations, you can follow these steps:

1. Determine process categories to make it easy for people to search for similar projects
2. Customize with fields like your global locations, value stream names, products, business units, and so forth, again to make it easy to search

3. Provide criteria to determine which projects can be checked as successful
4. Determine a standard problem-solving process
5. Provide an explanation and examples of hard and soft savings
6. Communicate which fields are mandatory

PROCESS STANDARDIZATION

If you have a global organization or have more than two or three locations within a country, all value streams must be standardized—something the top fast-food organizations have known for a very long time. Without standardization it is very difficult to share knowledge or transfer people between locations, another method of knowledge sharing. Without standardization, people also develop their own language and set of acronyms, which again provides hurdles for knowledge sharing. Process standardization provides very fast-paced knowledge sharing as the organization quickly learns and implements these otherwise hidden benefits and overcomes the "not invented here" syndrome.

Following are steps to take to get standardization into place:

1. Select a location with the best practices and results for a certain value stream.
2. Write a video script of these best practices, which will include lower-level metrics.
3. Hire a professional video company along with a narrator to shoot and speak on the best practices of the value stream.
4. Review and edit the video.
5. This video now becomes the current standard for this value stream.
6. Distribute the video to each of your locations.
   a. Each location forms a process-knowledgeable team to review the video and propose best practices with proven results that might not be in the current video. These are then sent back to the corporate team that is leading the standardization.
7. With all locations providing their feedback, a joint conference call with all of the location team leaders is held.
   a. This meeting is used to achieve consensus on which best practices should be added to the video.

8. The location where the first video was shot is then given some time to implement these best practices.
   a. If some best practices are capital intensive, the video crew will have to travel to the location where the best practice was realized.
   b. The video script must be updated to include this best practice.
9. Once the location is ready then the video is redone to add the new best practices.
   a. This video now becomes the new standard.
10. The new video is now redistributed to all of the applicable locations.
    a. They are now responsible for implementing and measuring against the best practice targets.
    b. A standard report is used for the locations to report with.
11. A monthly conference call is set up with the locations.
    a. The conference call provides support and further knowledge sharing of implementing the best practices.
    b. There must be at least a vice-president on the conference call.
12. A best practice validation process is set up with various experts of the value stream.
    a. The locations are informed they have a responsibility to send the corporate team any new best practices realized.
    b. With new best practices the video is once again updated and distributed for best practice implementation.
    c. This new video is now the current standard, and this process repeats itself whenever there is a new best practice realized.

I should also provide some description of a best practice:

- A new technology
- A process
- A method
- An interpersonal method

All of the above must have proven (over time) and measured results that exceed those at your other locations.

One concern people have about any type of standardization is it takes away people's creativity. Unless you are installing all robots (and even this type of standardization can be improved), that is not the intention. I promote that standardization is a continuous improvement process. The continuous improvement efforts continue and collaborate across the enterprise, which presents the other benefit of fast-paced knowledge sharing. Once an effort has produced a process result exceeding the effect of the standard, the solutions are validated through a standardized validation process, similar to Step 12 above. I have also found that people become motivated in a competitive way because if their improvement is accepted they now hold the "world record"!

SOCIAL MEDIA

Today, sharing knowledge and finding answers to questions proposed are happening at a pace we have never seen before; in fact it is growing exponentially. LinkedIn, Twitter, Facebook, Flickr, and other social media have exploded. I have found, however, that in some cases people do not want to divulge complete details, but they will provide a starting point of where to find more information. Today, companies are starting their own social media where any employee at any level can pose a question, submit an idea, start a category or tag, and provide answers. This does take some investment by the IT department, but it does not take very long to get up and running. Of course it is global, and you can even allow suppliers and customers to contribute to selected categories. Just think of the benefit of collaboration with your trusted customers and suppliers.

Organizational technical and process experts will soon become apparent not only to the organization but also to the trusted suppliers and customers. This creates a transparency benefit that people will appreciate and, maybe more importantly, provides a competitive edge as customers will now go to your organization for questions they need answered for a certain product or service.

The organizational social media works well because, for most workers, there is an organizational filter of knowledge sharing; in my experience, word-of-mouth ideas have only a 1% chance of being implemented, suggestion programs less than 12%, and even successful projects stored in databases are copied less than 20% of the time. Organizational social media is not very effective because the idea put forth is filtered by someone with authority or lack of exposure to many other people within the organization, or they have to cross departmental boundaries within an organization. Now I know some people are reading this and saying to themselves, this could create chaos within an organization. This is your paradigm getting in the way. For those of you with this paradigm, you have to question if you yourself feel empowered within your organization to share knowledge. In any case, most organizations already have their company as a group on LinkedIn, Twitter, and Facebook. These groups were started by their employees and are not managed by the executives of the company. It is interesting that employees are taking this initiative. We know that a fast-paced learning organization will stay ahead of the competitors, so we need to remove as many filters and roadblocks as possible. Everyone must be able to contribute their creative ideas at a fast pace.

Another benefit of developing your own organizational social network is in standardization through collaboration across functional groups. Each of these functional groups or departments has different cultures and even language, presenting barriers. The starting of social groups (groups are subject specific, like a certain technology, but the members are cross-functional) can rapidly increase standardization because of providing near-real-time updates and support of required implementation tasks. Not only can the project leader provide the updates, but so can other stakeholders like self-directed teams provide instantaneous updates with real data and general information.

If we go back to when Gutenberg invented the first printing press around 1439 in Europe, we can start to see the evolution of the knowledge-based economy and how the pace was increased. Gutenberg allowed the mass production of printed books, which made it economically viable. Information or knowledge sharing picked up to a new pace. More recently the Internet has significantly increased pace in knowledge sharing, beginning as the ARPANET in 1969, when a message was sent from a computer at the University of California–Los Angeles to a second piece of network at Stanford Research Institute. The Internet developed later and connections have grown rapidly since, with all kinds of more economical methods to connect gadgets like mobile phones and with more and more people sharing information. In the past only the elite few, like journalists and academics, published key information. With the social networks I have mentioned, anyone can publish information and for the most part it is unfiltered. The social networks have increased the pace of knowledge sharing once again.

IDC projects that the digital universe will reach 40 zettabytes by 2020, an amount that exceeds previous forecasts by 5 ZBs (Gantz and Reinsel, 2012).

Is the next step to increasing the speed of knowledge sharing a socially networked enterprise instead of a hierarchical organization? The department silos are barriers to a fast pace because of protecting what is required for their group to be successful within a hierarchical organization. If we take, for example, an organizationwide global implementation of Lean, within a hierarchical organization the global Lean strategy is filtered by first the division leaders, then by the location leaders, and then by the departments within these locations. These heads of divisions, locations, and departments might represent 1% of the organization's population. Typically it starts in a certain segment of the organization; in a Lean implementation it is typically manufacturing. So taking advantage of Lean benefits by the enterprise might be a 5 to 10 year endeavor; in fact it might not never happen. With a socially networked enterprise the global Lean strategy is communicated across the enterprise within days because of the network. The global Lean strategy would, from the beginning, include all segments of the enterprise: sales, HR, finance, IT, and others. Everyone would be clear on what their role would be, the skills required, their questions answered (removing any fear), who the experts are, what's in it for me, who the groups of expertise are, what the segment business targets are, and what is the leader involvement. Socially networked organizations are the next step in significantly increasing pace.

SUCCESSION PLANNING

Organizations are starting to realize the value of knowledge sharing, but too many organizations do not really understand how to effectively use it. Knowledge sharing can dramatically speed up an organization's ability to innovate and to solve problems. Think of the wasted waiting time and cost your people incur by waiting for experts from the supplier or from another part of the world, waiting for industry consultants, and waiting on internal departments. Having expertise dispersed among your people will provide a faster pace.

Knowledge sharing is not just about technology. Having the IT department start SharePoint is not the only element for knowledge sharing. Ask an expert where they uploaded a successful project and they usually will need to take some time to find it. Knowledge sharing is mainly about people interacting. The organization must provide these interaction opportunities through an internal social networking system and succession planning. There are also technology elements like SharePoint, but I will discuss the technology aspect in an upcoming blog.

HR plays the main role in succession planning, which is one of the central reasons they are a member of the Management Steering Team. Every organization has experts in a variety of disciplines, including the workers directly working in the value streams. There must be succession plans developed for knowledge sharing. Who are the successors and how do these successors learn from the experts? What motivated these experts? How did these experts learn? What was their learning process? These are the critical questions to be answered by the Management Steering Team, with HR taking the lead. Most organizations have succession plans. Unfortunately, they are reviewed once per year and then secretly stored. With knowledge sharing as an organization value, the succession plans now have a valued purpose of developing an effective and extensive network of experts.

The Management Steering Team is responsible for developing a knowledge-sharing map to ensure an effective succession plan. Here are the steps to develop this knowledge-sharing map:

- The key disciplines are aligned with strategy deployment.
- The experts are identified for each of these disciplines (internal social network will make these experts obvious).
- Determine how the current experts became experts, how they learned.
- Identify the successors.
- List the required resources and methods for learning.
- Develop metrics to determine effect, like waiting time for knowledge and associated costs.

The knowledge sharing map is then executed with the successors, and the experts are now mentors. For the last mapping step, here are some resources and methods HR can make use of to facilitate the learning process of a succession plan:

- Suppliers provide some of the teaching by coaching the successor at their facility.
- Attend industry conferences.
- Benchmark with other organizations who use similar technology or practices.

- Train the expert as a mentor for the successor.
- Develop a library of reference material for all disciplines.
- Schedule planned learning events at sister locations.
- Have an expert outline key project reviews from databases like SharePoint.
- Become a member of active forums on external and internal social networks.
- Provide contact information for industry experts in other organizations and your organization. Have the successor contact them at various intervals to understand certain subjects.
- Provide opportunities to teach about certain subjects and to speak at industry conferences.
- Work alongside other known experts.
- Become trained in TWI-JI and perform job breakdowns and then train other people.
- Submit an industry technical paper.

It is also the responsibility of the Management Steering Team to list the disciplines required to have succession plans. I remember working for an organization where everyone on one location's management team had a technical background. They had succession plans for their engineering roles but nothing for the people working directly in the value streams. Within their value streams they had a process that was extremely unique, so much so only a few people in the world had this knowledge and practiced it. However, their key customers very much valued it. It wasn't until the two operators who knew the process were into their 60s that they realized they needed to have a succession plan!

As with other strategic initiatives, visual management and Standard Work for Managers are key Lean Sigma programs to provide fast-paced progress. Post visual metrics in a common area of the Gemba. This visual management will be a matrix of the learning process resources and methods I listed above. The matrix will include the planned dates to have tasks completed, and any results and comments. It will be a joint responsibility of the expert mentor and successor to ensure the dates are met.

With the visual management in place, the Management Steering Team can design a checklist for the Standard Work for Managers program. The checklist will be used mainly by HR and the leader of the organization. They will use the tiered Gemba walk practice and also sometimes take the mentor and successor on the Gemba walk to review the results and progress of the plan.

Knowledge sharing will contribute to becoming fast paced, providing a competitive edge. And succession planning is one of the key elements.

BREAK DOWN THE SILOS

Organizational departmental silos are barriers to fast-paced knowledge sharing, standardization efforts, and change. Over time, departments develop their own cultures and work toward strongly protecting this culture. If the tribe within the department senses an attack on their culture, they will meet as a tribe and either determine how to appear in compliance with the threat or will not actively participate in what is being asked of them. Let me give you an example.

A geographical marketing division, with many years of successful sales resulting in a high contribution margin ratio, was asked to start developing the technical skills of their salespeople. It was explained to them from corporate marketing that customers were starting to request more total solutions than the basic product. The response from the geographical tribe was outrage: How could they question the long history of success? What we have been doing is very successful! If it ain't broke, then don't fix it. Their paradigm was developing salespeople who understood the product basics, aggressive and frequent visits, and customer entertainment. The tribe looked at this as an attack on their successful model; they didn't want to change to technical geeks who knew nothing of the customer! They were asked to present their plan for this new frontier. So as a tribe (head of the division, sales manager, and a few other top salespeople), they entered the conference room and presented nothing but the history of their success, thinking they would convince the CEO he was wrong in changing the approach. This disappointed the CEO, who replied "That was yesterday; you understand what the market is now starting to request, and we feel this is a trend that will only grow. I do not accept this proposal, so at the next conference I want you to present on how you are going to prepare and develop your people for this new trend." The tribe was outraged to the extent where one of the top salespeople stood up and said, "My customers buy from me and not from this company!" Unfortunately at the next conference the tribe presented almost the same proposal. Within two months the division manager was asked to retire, the sales manager left, and some of the other salespeople left the organization. They were all replaced, and the trend of customers wanting more solutions from the product continued to grow. The departmental silo in this case slowed the ability for the organization to react to an emerging trend.

I have even seen department visions completely separated, not aligned, with the organization's vision. One example is a department's vision of "Through the use of high-tech machinery we will become better than other departments," when the organization's vision was "Through synergy and teamwork we will achieve lower operational costs than our competitors." The department was focusing their resources on higher costs while the organization was clearly looking to be low cost. These conflicts and cultural differences are the barriers in achieving a fast-paced organization. These silos are not limited to the departments. Most organizations may have silos or cultural differences even within small groups of people when knowledge is not shared. With limited knowledge sharing, people's value within the organization is directly related to their unique knowledge. This can be seen as pockets of expertise which slows the organization down. Using the techniques of fast knowledge sharing will bring speed to the organization.

REFERENCE

Gantz, John, and David Reinsel. 2012. The Digital Universe in 2020: Big Data, Bigger Digital Shadows, and Biggest Growth in the Far East. *IDC iVIEW*. December. http://www.emc.com/leadership/digital-universe/iview/index.htm

5

Lean Sigma Tools

BRAINSTORMING

Brainstorming is a powerful yet underutilized Lean Sigma tool. It provides tremendous benefits for self-directed teams. When I am facilitating a team I use brainstorming at least two times during a one-hour meeting. Here are some of the main benefits:

- Generates new ideas
- Generates many ideas
- If brainstorming rules are maintained, then the usually quiet people on the team will participate
- Prevents getting stuck on the details instead of being creative
- Allows team members to consider every possible angle of a problem
- Allows team members to build on each other's creativity

Having the participation of the entire team, including the quiet people, is probably the most beneficial, as every team will have members who are willing to allow the others to provide all of the ideas. I once was facilitating a team in a service company with one team member who rarely commented on anything we were doing. Once we started using brainstorming and adhering to the rule that "you are not allowed to critique any ideas put forward until the rounds are complete," this very quiet team member commented that "you do not give us enough time to complete the process." I do not want to bore you with all of the steps taken after this, but the end result was that this simple statement led us to a worldwide yearly savings of $2.25 million, in hard savings, from an improved maintenance and safety program.

It is of course absolutely critical that the leader of the brainstorming session ensures the rules are adhered to by all members. Any deviation will take a lot away from the success of the session.

Brainstorming Rules

1. Write the problem at the top of the board.
2. Set a time limit (5 to 20 minutes).
3. Ask each member in rotation. Everyone takes part; members are allowed to say pass, but only once.
4. You are not allowed to critique any ideas put forward until the rounds are complete.
5. Let yourself go—wild ideas are welcome.
6. Quantity of ideas is the goal; continue the rounds until ideas are exhausted.
7. Write everything down.
8. Spelling is not important.
9. Group ideas.

Next, determine which ideas are selected:

1. Number the ideas in the brainstorming list.
2. The number of votes per person is one-third of the total number of ideas in the brainstorming list. For example, if there were 18 ideas, each person would have 6 votes.
3. Everyone places their votes according to the idea number on a piece of paper.
4. The person conducting the brainstorming session collects all of the papers.
5. The ideas with the most votes are selected.

5S

Sort, Straighten, Sweep, Standardize, Sustain—5S not only prevents unnecessary stops and provides a clean work environment but it also instills process discipline, reduces product defects, and reduces cost. If you do not have 5S in place today or if your 5S program has completely lost its momentum, it is time to take a white glove tour of the shop floor. I am sure you will find damaged raw materials, empty consumable bins, disgusting amounts of dust and dirt on top of your valuable production machines, dust and dirt on anything with a vent including machines and computers, broken or missing tools and measuring devices, material piled up on production machines, outdated postings, people having to wait or walk too long for a quality measurement, WIP with no apparent destination, and more. You probably also have to have production shut down so that operators can clean because a customer or your boss is coming for a visit. 5S is simple but powerful for the benefits you receive.

The definitions are as follows:

Sort: Proper arrangement. Red tag and remove from the workplace any item that is not needed for current production or the current day's assignment and anything like wooden pallets that collect dirt. Keep only what you need today. Add anything that is needed but not there.

Straighten: Orderliness. Arrange items so they are easy to use. Mark and label these items so they are easy to find and put away. Have a designated place for everything. Store everything close to the application of use.

Sweep: Cleanliness. Sweep the floors, continuously clean equipment, paint if necessary, and in general, make sure everything in the plant and offices stays clean, continuously.

Standardize: Cleaned up. Standardize and maintain the use of sort, straighten, and sweep.

Sustain: Discipline. Practice and repeat these procedures until they become a way of life throughout the entire business.

Sort

Sorting starts with red tagging the items, as described above, with an actual tag like the one in Figure 5.1. This removes any items that present clutter in the work cell, which can cause defects, make communication more difficult, take up valuable space, and cause wasted time in searching for required tools and materials. It also brings required tools and materials closer to actual location of use, saving valuable steps and time. The red tag is filled out and tagged to such items. The red tagged items are brought to a taped-off area of the cell and will remain there for approximately one

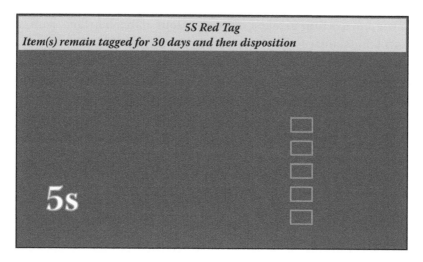

FIGURE 5.1
5S red tag. (From Robert Baird.)

month. This gives other people an opportunity to review the deposition of these items and if justified can change the deposition. For example, a machine rail that has been stored in a desk drawer for six months has been red tagged and is now stored in the red tag area. A process engineer happens by and reviews the items in the red tag area and notices the rail, which is a part needed in another cell. The process engineer notices the deposition is to be scrapped. The process team coach for the cell is notified, and the cost of the rail has now been saved and properly utilized.

Straighten

In the straighten step, items like, tools, consumables, measuring equipment, computers, and office supplies are moved closer to where they are actually used in the cell (see Figures 5.2 and 5.3). For these items to be properly placed and categorized, they need to have proper containers and locations that are conducive to easily seeing the reorder points, identifying when something is missing or is in the wrong location, and reducing wasted motion. Everything that is used in the cell by the team members must have a visual label. Visual is being able to see at a few paces away. Items like those in the list above all receive a label. Items like consumables must have calculated minimum and maximum amounts. With this in place you can even go a step further by establishing a program with your supplier to have them ensure you always have the correct amount of

FIGURE 5.2
Daily tools used on a machine. (From Robert Baird.)

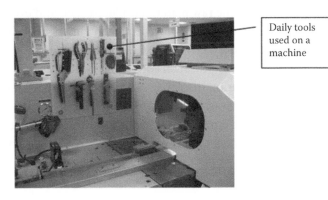

FIGURE 5.3
Daily tools used on Machine 2. (From Robert Baird.)

consumables (see Figure 5.4). Production machines should also be labeled by the language of the shop floor. If management identifies a machine by Machine 00998543 but the team members know it as Hot Stamp 3, then label it as Hot Stamp 3. Measuring equipment needs to be labeled with the same concept, but also add what the device measures. If the device measures die thickness, add this to the label. If the shop floor is using Kanbans

FIGURE 5.4
5S for consumables. (From Robert Baird.)

for raw materials, they must also receive a label and a color-coded tape on the floor to identify where the raw material is placed each and every time. All information boards must be reviewed for content. If nobody really looks at it, if documents currently posted are out of date, and if documents are attached by tape or pins, consider using electronic boards with LCD TV displays. Information displayed can then be controlled, owned, and updated without many wasted steps. Metric boards must have a visual means of ensuring data are current and graphs are either showing a good or bad trend by adding a green circle or red X to the corner of each graph.

Sweep

The cell area and machines must shine. If the floor is in a condition of wear, replace the floor; the cell must shine. If the machines are old with chips and missing items, bring them back to a new-looking condition. People have a difficult time cleaning something that remains substandard even when all the cleaning is complete. When cleaning, use a vacuum as much as possible; sweeping with a broom will generate airborne dirt, only to cause defects in your product. Purchasing a central vacuum system with hose connections to cover the cell is well worth the investment. Generate standardized work instructions to provide a procedure with the most effective methods for cleaning the complete cell. Determine the cycle time of each task and record it in the document. Remember, all cleaning equipment has to be 5S'd in order to provide efficiency in the cleaning process.

Standardize

Without an agreed-upon standardized schedule by the shop floor team and their process team coach, the cleaning methods and process will fade away. The agreed-upon schedule must determine when each cleaning method is started and stopped (cycle time) and by whom. The cleaning steps should be staggered to minimize process flow interruptions. The schedule must also be coordinated with other shifts. It would be a waste for the first shift to clean the floor at the end of their shift and second shift to clean it at the beginning of their shift. The reason for the agreed-upon schedule is to prevent the process team coach from stopping cleaning when production needs "must" get out and so that the standardized work can be audited. The audit would include a validation if the team members are starting at the time the schedule says. I can assure you, this discipline takes some time to get as culture. Once agreed upon, the schedule is posted in the cell, and nonconformances with action plans are also posted.

Sustain

The sustain step must be used in any change we make in a process. It is a daily audit of the 5S standardized work. Once 5S has been implemented, a T-card is added to the board and daily audits begin. The daily audits never stop, because 5S, like many of the Lean Sigma tools, is a continuous improvement program. This is how we surpass average results and obtain world-class results.

PROCESS STEPS TO ZERO QUALITY DISCOVERY

Zero Quality Discovery is a very powerful and effective methodology to identify both root causes of quality issues and critical inputs to control the process. The other benefit is that, like most Kaizen events, it is conducted during live production runs. When conducted properly, it is not uncommon to realize zero defects within two or three production runs. The preparation for the actual event is a key part of the methodology and will determine success or failure. The methodology is presented below.

Purpose

1. To determine and eradicate root causes of chronic and critical defects
2. To determine if critical process inputs are known and used as standard work

Required Resources

- Zero Defect Team
- Quality Manager and/or Quality Engineer
- Lean Sigma Black or Green Belt
- Quality Inspector
- Production Supervisor
- 1–2 Machine Operators
- Maintenance
- 3–4 Hours of Live Production Time

Accompanying Lean Sigma Tools (Preproduction Run)

- SIPOC (Suppliers, Inputs, Process, Outputs, Customers) diagram
- Cause and Effect
- Standard Work
- Flowchart
- Pareto
- Brainstorming
- Process Capability
- Kaizen

Process Steps

1. Management Steering Team assigns project charter to a project leader and recommends team members. Project charter includes the following:
   a. Project objectives with deliverables
   b. Project metrics with targets

   c. Project time lines with milestones of which will be used as project reporting points to the Management Steering Team

   d. Project budget if required

   e. Expected cost savings

   f. Project scope

2. Project team holds first meeting:

   a. Reviews project charter

   b. Determines schedule of milestones

   c. Assigns roles and responsibilities

3. Project team reviews current historical data of process defects.

4. Project team generates a SIPOC diagram to determine:

   a. Who the suppliers are of this process and what relationship they have in process defects

   b. What the critical inputs are for this process and what their standard measurements are when the process is running

   c. What the expected outputs are of this process (defects, production rates, and cycle time)

   d. Who the customers are of this process

5. Process team generates a process defect Pareto chart.

6. Process team generates a cause and effect diagram for each of the defects within the scope of this project. The completed cause and effect diagrams are used the day of running the zero defect batch to ensure the critical causes are not present.

7. Process team generates some possible solutions for the main causes.

8. Process team reviews the current maintenance condition of the machine:

   a. Review maintenance records (maintenance database)

   b. Review recent preventive maintenance performed

   c. Review autonomous maintenance tasks performed by the operator

   d. Q&A with the machine operators to determine their view on the machine maintenance condition

   e. Q&A with the maintenance technicians to determine their view on the machine maintenance condition

9. Project team determines a maintenance schedule to repair anything that is considered critical to achieving output objective.

10. Project team works with Planning and determines possible work orders for zero defect batch Kaizen. Order size is limited to no more than a two hour production run.

11. Check sheet templates are prepared for capturing the critical outputs of the process.
12. Project team determines date and time to run the first zero defect batch.
13. The day before running the zero defect batch, the project team communicates the project objective and zero defect batch steps to the machine operators involved, the supervisors, department managers, and next operation.
14. Steps to complete the day of executing the zero defect batch:
    a. Everyone on the project team is reminded of their responsibilities.
    b. All raw materials are inspected for defects.
    c. Work order information is reviewed for any mistakes.
    d. All required information, files, and materials are ensured to be at the operation on time.
    e. The shop floor is reminded of running the process of zero defect. This is completed in a shop floor meeting with the project team and the people who normally run the process.
    f. Process team and machine operators review the current standardized work documents for this process.
    g. Machine operator and other required support personnel perform the required machine setup.
        i. Project team is observing, noting any measurements taken by operator, and further improvements are also noted.
        ii. Setup parameters are compared to standardized documents and corrected if required.
        iii. Critical input parameters are measured, noted, and corrected if required. This could be machine settings, material specifications, and reference standards. These critical input parameters might not be measured today but must be done for running a zero defect batch. For example, at embedding it might not be normal to measure and inspect the cavity, but because it is a critical input parameter it must be completed even if it is completed through a large sample.
    h. Autonomous maintenance is reviewed and performed.
        i. Machine is cleaned within machine areas that could have a negative impact on the product quality.
        ii. Adjustments and setups of any sensors are made.
        iii. Critical areas are checked for tightness.
    i. Any variance from standardized work documents is noted.

15. Actual running of work order is started:
    a. Quality inspectors are inspecting 100% of the product at the output side of the machine. The machine must be run at a speed for unit-by-unit inspection or there must be enough inspectors to keep up with the machine speed.
    b. As soon as a defect is identified, the operation must be stopped:
        i. The complete team inspects the machine or material for the possible root cause of the defect.
        ii. All inspections and conditions are noted.
        iii. Causes found are noted and corrected.
    c. All defects are recorded by amount and category.
16. Project team generates summary report with the following sections:
    a. Project objective
    b. Names of all people involved and their positions
    c. Material references
    d. Process flowchart
    e. Critical input parameters and their measurements
    f. Step-by-step procedure
    g. Defect summary
    h. Causes of defects
    i. Conclusions
    j. Recommendations for process improvement and next zero defect batch run
17. Next zero defect batch Kaizen for the same process is planned.
    a. Average results are not accepted; only zero defects.
18. Standardized work documents (TWI) are generated for the newly found key points preventing defects.
19. Operators are trained with the new TWI-JI.

TPM

Total Productive Maintenance is a critical shop floor tool, the key word being *total*. Along with 5S it is a must program in implementing Lean. Without being able to run your machines on demand, Lean tools such as one-piece flow and Just-in-Time (JIT) do not have much of a chance of being successful.

I cannot emphasize enough the importance of getting everyone involved. Yes, total includes management. In most organizations, maintenance responsibility is delegated down to a maintenance department or other people working on our valuable products.

About 70% to 80% of the improvements come from three initiatives: Autonomous Maintenance (discussed in Chapter 3), Preventive Maintenance, and Predictive Maintenance. These are the initiatives I will be discussing for successful TPM implementation.

The Management Steering Team needs to be supporting, teaching, and promoting (STP) the TPM culture to keep your equipment assets running on demand. This does not mean they have to be directly involved in the repairs and preventive methods. They must, though, be involved in meeting with the key people on the condition of these key assets. They must provide an environment conducive to shop floor empowerment. They must provide the time and resources required to train so that empowerment can take place. They must provide an organization to support TPM implementation. They do all of this during their daily Gemba walk. Instrumental to the success are the self-directed teams. The teams need to provide metrics such as percentage of preventive maintenance (PMs) complete, percentage of TPM tasks completed by maintenance technician, and number of maintenance interventions per million parts produced. These metrics are both visually managed and presented in a weekly meeting and then discussed for improvement and for the effect TPM is having on the higher level objectives. A connection must exist to the top-level objectives.

Our weekly meeting agenda might look like this:

1. Summary of last week's minutes and actions
2. Review of metrics
3. Review of a machine PM task list
4. Determine actions for small repair list identified by team
5. Open discussion

In other Lean manufacturing books, you might have read recommendations for overall equipment effectiveness (OEE) as the overall metric in determining machine efficiency. Although it is a very good metric, I find that the input time involves too much time from the self-directed teams and so they do not use it because it is too complicated. It ends up being a waste and the data are inaccurate. One of the suggested metrics above is percentage of PMs complete; this is easily understood and is on the input side of our process. No, it is not going to measure the output metrics of Availability, Performance, and Quality, but which machine input has the most impact on these outputs. I always prefer thinking along the lines of an Input-Process-Output (IPO) diagram when determining a metric the self-directed teams can directly control and improve. So, percent of PMs complete is the most practical in this case for having an effect on machine breakdowns and quality. You might also be asking why the self-directed teams would be the owners of this metric; should it not be the maintenance department? The answer is that the self-directed teams are the owners of their part of the process and the assets within. Also, some organizations decide to have the maintenance techs as members of the self-directed teams.

Most companies cannot or are not even close to achieving 100% PMs complete and also do not have a PM continuous improvement program. We easily achieved 100% PMs complete within six months. We did this by our total involvement. The machines now run on demand, and we have reduced machine cycle costs by 179%. The people involved in this achievement were the Management Steering Team, maintenance techs, and the self-directed teams. The Management Steering Team reinforced that completion of PMs is more important than production needs. This was difficult in the beginning, and many communication follow-ups were completed before people were convinced this was the case. The maintenance techs, process engineer, and self-directed teams started by developing standardized work procedures for both the PM process and the PM tasks of what had to be completed at the time of machine intervention. This was done for every machine type. Mean time between failures (MTBF) and machine capability were used by management as the metrics to determine the success and the control of the PM improvement project.

One realization from reviewing our PM standardized task list was a very small percentage of replacement parts. This was because the inventory parts list was based on the PM schedule and not on a feeling someone had. Most of the PM standardized task list was lubrication, sensor adjustment, identification of worn parts, checking for wear, and cleaning. Although all-important, performing these tasks alone was not going to get us to our goal of machines running on demand. We all know that parts wear out, whether they are moving or stationary in some cases. So, the next step was implementing the Predictive Maintenance program. We needed to determine the life cycle of critical parts and sensors. This life cycle can simply be accomplished through estimation from the self-directed teams and maintenance techs, as a start. You might be surprised by how accurate this is. To validate, though, there should be an ongoing tracking program using logbooks, check sheets, or more sophisticated software. The measurement here must be number of machine cycles or units produced to determine part life cycle. Too many people determine this by time intervals, and of course, this can be very inaccurate with low machine availability intervals. To accomplish the machine cycle count there must be a counter for each machine, which is recorded into the maintenance software each time there is a machine intervention. For example, when one of the critical parts is installed, the unit count is recorded; when it fails, the unit count is recorded again, and the difference is the life cycle of the part. There is then a safety factor of ±10% taken from this result. These measurements continue until a distribution of failure is established as the part will not always fail at an exact count every time, but a high percentage of the time you will be making the correct decision in replacing the part before catastrophic failure.

With the critical part life cycle determined, the PM is now very similar to the PM standardized work determined by your car manufacturer. At various cycles (mileage) the car manufacturer manual provides a table to replace various critical parts.

Another metric that we use to determine the performance of our machines is the following:

$$\text{Performance Gap} = \text{Process Capability} - \text{Machine Capability}$$

Machine capability is defined as short-term runs that are able to determine the optimum outputs. If your maintenance staff is not calculating the performance gap, they are not involved with the TPM concept.

There are many reasons to determine the capability of our machines:

1. To set a baseline of performance. This baseline can then be used to determine a performance gap for the scope of the improvement process.
2. To determine if the current performance (new or refurbished machine from supplier) is acceptable according to the machine capability database. Machine speed and product yield do not tell the complete story.

3. To determine if a reduction in the performance gap has been accomplished after years of production or at any point after the arrival of the machine.

To determine the capability of the machine:

1. Reduce the variation of machine inputs by using the same operator, same material, and same machine settings while performing the test.
2. Run the machine and measure the machine outputs; machine must be run at the expected process speed. A minimum of 50 measurements is required, but 100 is optimum; depends on economics sometimes.
3. Determine that the Cm and Cmk from the output measurements and Cmk are the same formulas for Cp and Cpk, respectively. Signifying Cm and Cmk indicates a machine capability, not a process capability

The histogram in Figure 5.5 is the capability of the die bonding machine for the output of the X position. We can see that the Cm value is 2.298 and the Cmk value is 1.993 (the figure shows Cp and Cpk, but this is the default of the software). We now have a baseline that Production can compare to. To complete this comparison, Production would determine their Cp and Cpk values (process capability, not machine capability) by measuring the machine outputs with all of the production input variation being realized. The difference between the process capability and machine capability becomes the performance gap, so:

Performance Gap = Process Capability (Cp) – Machine Capability (Cm)

This performance gap can now be an indicator of optimized or deoptimized performance for the future of this machine.

Machine Green Zones

An argument I was confronted with had to do with both the safety of the machine operator and how the machine operator could possibly damage the machine if they adjusted certain input parameters. The maintenance technician would present the example of how the *Titanic* would sink if we allowed the machine operator to make machine adjustments. Well, in this case, the *Titanic* would have completed the voyage, and in many cases

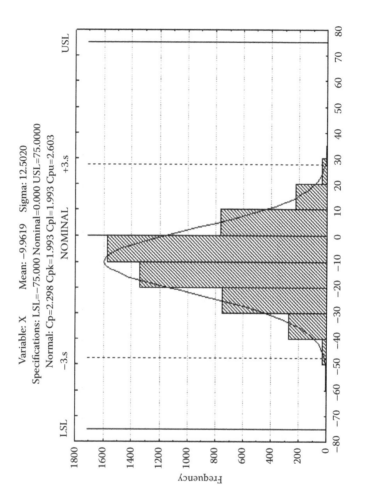

FIGURE 5.5
Die bonder distribution. (From Robert Baird.)

we learned that the machine operators were much more capable than we had assumed.

One of the arguments I would present is that if some of the operators were completing these adjustments, what was preventing the other machine operators: motivation? empowerment? Most likely the latter. Another reason for machine operator participation in maintenance is the time involved to complete a certain machine repair or adjustment. One time, when installing a beneficial maintenance software program, we realized that 80% of our maintenance requests were 15 minutes or less, but the response time (time it took for the maintenance technician to get to the machine) was more than double the repair time. It was now obvious to have the machine operator complete these types of repairs or adjustments.

To facilitate the operators in doing machine adjustments safely and without damage to the machine or product, we installed what we called Green Zones. The idea was to mark all gauges, adjustment knobs, and anything like railings that had to be adjusted throughout the course of the shift or a setup with green, indicating the safe operating zone. If the Operator made the adjustment within the zone and it had no effect, then a maintenance request was generated. We also looked at maintenance requests as an opportunity to teach and learn. The maintenance people were empowered to provide training on what they determined was something the operator should be capable of learning. They would not do this on the spot but would later use the TWI-JI training program to validate, prepare the job breakdown, and then actually perform the training.

Measuring the Maintenance Technician Involvement in TPM

TPM is not a tool the shop floor understands and adopts immediately. The maintenance technician quite simply wants to repair the machine; it has been their objective for years. Here is a simple program we used to obtain the involvement of each and every Maintenance Technician.

Process for Implementing TPM

1. Select a project implementation team. Must be made up of the following:
   a. Lean coordinator
   b. 2 to 3 maintenance technicians

    c. If there is a maintenance manager please include

    d. 1 shop floor supervisor

2. Outline a meeting agenda to inform all maintenance technicians of the following:

    a. The objective of implementing a TPM program

    b. The benefits of a strong TPM program

    c. How they will be involved

    d. General training they will receive to understand Autonomous Maintenance, Predictive Maintenance, and Preventive Maintenance

    e. The associated shop floor supervisors must attend these meetings.

3. Outline a one hour training program. The training materials and execution can be completed by the Lean department or anyone who does some research on TPM in general. The training must include the concepts of Autonomous Maintenance (operators completing small repairs and adjustments to their machines), Predictive Maintenance, and Preventive Maintenance.

4. Communicate the training date and time to the target maintenance technicians and confirm attendance. Please CC the associated shop floor supervisors, department managers, maintenance manager, and plant manager.

5. Develop a checklist for the maintenance technicians' usage in the day-to-day recording of their daily tasks when working at a machine (Table 5.1).

6. Develop a monthly summary report (Table 5.2; note that example data are included in the table). Assign a TPM team member to summarize the data from the daily Table 5.1 to Table 5.2. The goal is to achieve greater than 90% TPM tasks. Of course you will never get there if most of the tasks completed are repairs.

7. Implementation team decides on which Autonomous Maintenance task they will train on for the current month. A different Autonomous Maintenance task is selected for each month. The maintenance technician is assigned to ensure that each operator receives the training and is competent in the ability to perform the task.

8. For each Autonomous Maintenance task selected, Standard Work instructions are developed using TWI-JI methods.

9. The program is started by the shop floor supervisor instructing the maintenance technician(s) to start using the daily checklist (Table 5.1). Direction is also given to provide the Autonomous Maintenance task training to the self-directed team members throughout the month.

TABLE 5.1

TPM Checklist

	Total Productive Maintenance (TPM)		
Date	**Task Type (Please check one)**	**Task Description**	**Maint. Tech. Initial**
	Corrective Maintenance Predictive Maintenance Preventive Maintenance Repair		
	Corrective Maintenance Predictive Maintenance Preventive Maintenance Repair		
	Corrective Maintenance Predictive Maintenance Preventive Maintenance Repair		

TABLE 5.2

TPM Daily Summary Table

Week	**TPM Tasks Completed**	**Repair Tasks**	**Total Tasks**	**% TPM Total**
22	4	18	22	18.18%
23	6	16	22	27.27%
24	4	14	18	22.22%
25	5	11	16	31.25%
26	6	11	17	35.29%
			Avg.	26.84%

They are also instructed on ensuring a percentage of Predictive and Preventive Maintenance tasks are completed.

10. Table 5.2 is visually managed on the TPM metric board.
11. The Management Steering Team is informed of the start date and includes the review of the progress in their daily Gemba walk.
12. At the end of the week the shop floor supervisor summarizes the data from Table 5.1 into Table 5.2. Table 5.2 provides the monthly summary.
13. After the first month of implementation the maintenance technician is assigned a single machine to start measuring MTBF. This is recorded and graphed into a shared spreadsheet for the Management Steering Team to review.

VALUE STREAM MAPPING

Value stream mapping (VSM) is one of the first key methods to use when starting a Lean transformation. It is used to first determine your current flow of key products and then to develop a desired or future state flow of these products. It is similar to putting together a process flow map, but it encompasses the complete process, starting with the customer requirements and working back to all of the process suppliers. It also provides all of the key input and output metrics. The success of this tool arises mainly from the establishment of a diverse team including key people from each part of the process. The second key method used is to have the team go to the Gemba and record actual data and observations. This not only provides an actual picture of the current steps but also additional knowledge for all of the team members. The value stream identifies all of the steps, both value-add and non-value-add. The value-add steps are all of the steps or actions within the process that actually transform the product or service the customer actually pays for. The non-value-add steps are steps like inspection, transportation, tool change, maintenance, and others—all steps currently required but that the customer is not willing to pay for. You can even see non-value-add steps or actions within a production machine like long transportation time to the next value-add station. Here are the main steps involved in VSM:

1. Identify each of the value streams (product flow). VSM is completed for each of your value streams, so the first step is to identify them. If you are already in manufacturing or office cells, this should be straightforward. You should start with products or services for your key customers or a customer you must improve for. Targets are set by the Management Steering Team related to the single-focus strategy. During the "kick-off" meeting the Management Steering Team promotes the objective of the VSM exercise.

2. Establish the current state (flow path). Start with the customer requirements, how much volume per month, lead time, on-time delivery, and quality levels. Calculate the Takt time. Enter all of the key customer data on the VSM. If the team is able to actually have the customer join this beginning of the VSM, it is obviously very valuable. The team will get nonbiased and factual information.

Remember, we start the VSM by gathering the customer requirements, so if the information is biased (as it can be by salespeople) or not correct, the future state is not going to provide the organization with much value. After the customer requirements are recorded, we then start working back to the steps of the process, including required process information. For each step we record the following:

- Cycle time
- Work in progress (WIP)
- Changeover time
- Utilization (this is the actual time the operation runs ÷ time available)
- Batch size
- Number of operators
- Yield loss
- Instantaneous run rate
- Value-add time and non-value-add time
- Transportation time and distance
- Square footage required

Below each of the VSM process step boxes, draw a data box and fill in the above information. If there are other flows, you can draw the VSM process step boxes above the main stream steps. Once you have all of the process steps and information on the map, the next step is to record the levels of WIP between each step and insert the VSM inventory symbols. This includes the backlog of orders that are in queue for the first operation. The levels of WIP are recorded below the symbols. The next step to record is how often raw materials are delivered to the location and how much.

3. Confirm the flow and data you have recorded on the map. Now is the time to actually go to the Gemba and validate the data and flow you have recorded on the map. This is critical as people's perceptions (even the experts on your team) can be very different from what is actually happening. The team starts at the end of the process at the warehouse and walks back through to where the raw materials are delivered. Information flow is also checked along the way. It is very important to get an actual count of the WIP as the total lead time (in days) is calculated for each inventory triangle by inventory

quantity ÷ the daily customer requirement. All are summed together. Another method is to take the total inventory quantity ÷ the actual production rate and then sum together; this is more accurate. At the end of the process map, record this process lead time, the total value-add time, the non-value-add time, total WIP, and the overall First Pass Yield.

4. Complete the current state map. The team can now connect all of the steps with arrows for flow direction. For push flow movement the arrow is striped. Information flow is one line and digital flow is a one-line flow with a hook in it. At the end of the process map, record this process lead time, the total value-add time, the non-value-add time, total WIP, and the overall First Pass Yield. The meeting wall now displays the current value stream. Now it is time for the team to reflect.

5. Generate the future state map. This is the step where the team needs to be creative and look at every opportunity to meet the targets given to them by the Management Steering Team. Review the non-value-add steps, like inspection; can they be eliminated; can they be combined into one step or completed by fewer people; or at a minimum can they be shortened or variation removed? Look at where you have utilization too high or low, a bottleneck, or improvements required. Look at where the production system can go to a pull system or, if multiple products are produced, a pull supermarket system. Look at where cells can go into place to eventually achieve one-piece flow. Look especially at where WIP can be lowered. Look at where transportation waste can be minimized with a new layout. Look at where there are excessive changeover times and reduce. Look at deliveries; can they be combined or done for less cost. Basically challenge everything! The plan should be divided into three time lines:, what solutions can be completed in 30 days, what can be completed in 90 days, and what can be completed in 120 days (you might have to extend over 120 days, but try not to as you are a fast-paced organization). Solutions are recorded in the VSM Kaizen bursts and pasted on the future state map. All changes to flow and proposed data are now put onto a new map called the Future State Map. The team stays together and meets once per week until the future state is realized. The team must understand that the improvements do not stop with the future state realized. At the point of realization the VSM team starts another current state.

STATISTICAL PROCESS CONTROL

Statistical process control (SPC) is a tool used to monitor and control a process. If used correctly it provides protection for your customers by being able to demonstrate patterns where, if not corrected, products will be nonconforming. SPC was pioneered by Walter A. Shewhart in the early 1920s. Shewart also developed the control chart and concept of the state of statistical control. When people think of SPC, they usually limit it to control charts, but the key tools used in SPC include control charts, a focus on continuous improvement, and the design of experiments.

SPC does help if, like any tool, used correctly. There are a number of errors made by people implementing SPC for the first time:

- Control limits are calculated when the process is out of control
- The wrong control chart is used (Xbar-R versus Xbar-S, for example)
- Customer specifications are used instead of the control limits
- Used with nonnormal data
- Control limits changed when there has been no process change

These are some of the reasons why people do not see the value in this tool. With the correct use of a control chart and with the process in control, SPC will provide a predictable manner that is valuable in itself. If the process data being plotted follow a normal distribution, any of seven data patterns will tell us the process is out of control. Each of these patterns will occur less than 1% of the time, which gives us confidence the process is out of control. So with the correct application and understanding, SPC will provide protection for customers.

The Key Steps in Starting SPC

1. Determine the software you will use for control charts and design of experiments (DOE).
2. Start with process variable data (something measureable like heat, pressure, length, etc.).
3. Start data collection plan. Collect, at minimum, 25 data points or subgroups.
4. Determine if data fit the normal distribution. If not, a DOE might have to be performed to understand where some of the variation is coming from.

5. With the data normally distributed and the process stable, calculate the control limits. The importance of this step is overlooked by many people, but it is critical to identify "special causes." Shewhart (1930) determined that all processes display variation, and that some of this variation comes from the standards of running the process, or *common cause variation*, and some comes from uncontrolled conditions that are not present at all times, or *special causes*. So if the data used to calculate the control limits partially consist of special causes, the control limits will not be calculated correctly and they will not be able to correctly identify these causes.

6. Determine which of the seven out-of-control patterns will be the most beneficial by identifying the characteristic you are concerned with. For example, a single data point going outside of one of the control limits might not be indicative of a blade losing its sharpness, but seven constant data points increasing is more indicative.

7. Start the training of SPC for the workers. Keep it simple!

8. Include the review of this control chart in the manager's daily Gemba walk. I think you will find that the initial SPC training is not enough for the workers to fully understand the data collection requirements, how to identify the out-of-control condition, and what to do. The daily Gemba walk can provide the support, teaching, and promoting required for the correct application of this valuable tool.

9. Spread SPC to other parts of the process.

10. Use SPC to continuously improve process variation and thus process capability. Elimination of sources of variation is best addressed with Standard Work, error proofing, or change in the process itself.

SINGLE-MINUTE EXCHANGE OF DIES

Reducing changeover time can make a significant contribution toward lead time improvement and capacity improvement. I have seen where Single-Minute Exchange of Dies (SMED) improvements avoided significant capital investments because of production time gained. Every organization has seen a change toward smaller order size and customization of these orders. Gone are the large batch sizes of yesterday. To become efficient, every organization will need to reduce the non-value-add time of setting up a process step.

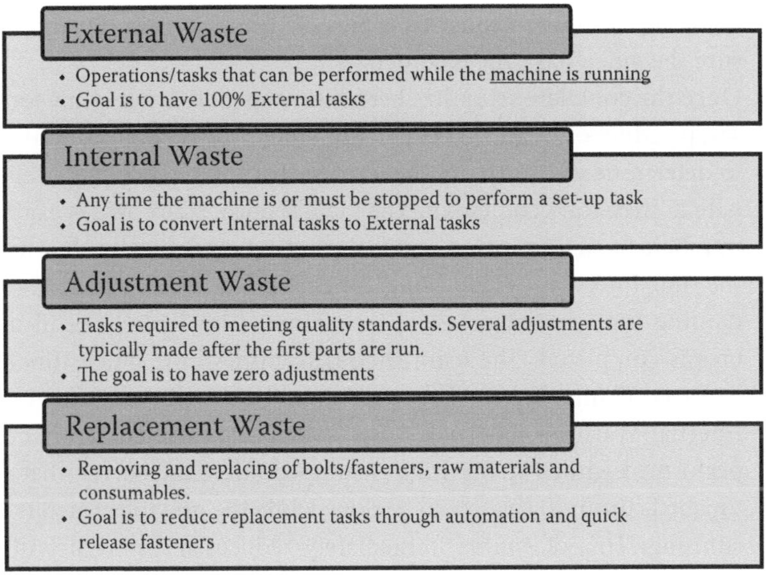

External Waste
- Operations/tasks that can be performed while the <u>machine is running</u>
- Goal is to have 100% External tasks

Internal Waste
- Any time the machine is or must be stopped to perform a set-up task
- Goal is to convert Internal tasks to External tasks

Adjustment Waste
- Tasks required to meeting quality standards. Several adjustments are typically made after the first parts are run.
- The goal is to have zero adjustments

Replacement Waste
- Removing and replacing of bolts/fasteners, raw materials and consumables.
- Goal is to reduce replacement tasks through automation and quick release fasteners

FIGURE 5.6
SMED categories. (From Robert Baird.)

SMED breaks the setup process down into four waste categories: External, Internal, Adjustment, and Replacement (see Figure 5.6).

During the SMED process a team identifies and analyzes these wastes. These wastes are then either transformed or removed completely.

SMED Implementation Steps

1. Form a diverse improvement team made up of mainly workers within the process, engineer, and team facilitator.
2. Determine metrics to indicate the success of the project:
   a. Setup time
   b. Capacity increase
3. Start with a process that has the longest setup time.
4. Videotape the complete setup process, from the time the machine stops producing the last unit of the previous order to the time the next order is continuously running. Defining the start and stop is important; for example, some people stop the time when the first unit is starting to run, but when the first unit hit the output side was it defect free? If not, there remains Adjustment Waste. Do not turn the video off if someone has to leave the setup to retrieve tools,

material, or information; this is part of the setup time. Lastly, make sure the video has time recorded.

5. Once the complete setup has been videotaped, it is now time for the setup reduction team to review the video, from beginning to end, to determine where any of the four wastes can be reduced or eliminated. They start completing the table (Figure 5.7 provides example numbers from a setup for an offset printing machine) by identifying the time intervals (Video Time) for each of the waste categories and naming the associated tasks (Task Time). The Task Time in minutes is completed. The team then determines how much time can be saved (Time Saved). The team should be looking at converting Internal Waste to External Waste—what tasks are currently being performed while the machine is stopped (after the current batch is finished running) that could be done while the previous batch is still running. This of course immediately reduces machine downtime because of setup. The team then provides a rating of the potential of each task to be reduced or eliminated by applying a score from 1 to 5, 1 being the lowest potential. Time Reduction is the potential to save time, and Opportunity is how easily we can design or change something to provide the time reduction. These two numbers are

Task #	Video Time	Task Time (min)	Time Saved (min)	Task	Time Reduction Rating	Opportunity Rating	Total Score
	**External Waste** *(machine running)*
1	1:33 – 1:34	1	0	Check with Maintenance	1	1	1
2	1:34 – 1:39	5	4	Getting setup tools	4	4	16
	Sub Total	6	4				
	**Internal Waste** *(machine stopped)*
5	1:48 – 1:49	1	0	Plate bending	2	4	8
6	1:49 – 2:06	15	10	Load inks	4	2	8
	Sub Total	16	10				
	**Replacement Waste**
8	1:41 – 1:51	10	8	Load book	5	5	25
9	2:05 – 2:15	10	5	Load core for job	3	3	9
	Sub Total	20	13				
	**Adjustment Waste**
12	1:54 – 2:04	10	10	Adjust ink density	5	3	15
13	1:55 – 2:01	6	6	Adjust registration	5	4	20
	Sub Total	16	16				
	Total	58	43				
	Time Reduced To:	15					

FIGURE 5.7

SMED video review datasheet. (From Robert Baird.)

then multiplied together, and the result is stated in the total score. This score provides the team with a prioritized list of where to focus their initial efforts toward improvement. The Task Time and Time Saved columns are then subtotaled for each of the waste categories to indicate which ones will provide the most time savings. Finally, all of the subtotals are summed to a total, and the Time Saved total is subtracted from the Task Time total to predict what the new setup time should be.

6. The team can now start putting the solutions together and implement.
7. Videotape the setup process again to confirm the predicted setup time.
8. This process is repeated again and again until the required setup time has been reached.

REFERENCE

Shewhart, Walter. 1930. Economic Quality Control of Manufactured Product, *Bell System Technical Journal,* IX, No. 2 (April, 1930): 364-389.

Glossary of Terms

5S: Sort, Straighten, Sweep, Standardize, Sustain. An efficiency and quality method used to ensure prevention of having to wait for something that is missing or damaged at the point of use. It is also a quality method because one of the effects, by removing items that are not needed, is a cleaner workspace.

Continuous Flow: Producing one piece at a time, with each item passed immediately from one process step to the next without queue.

DFSS: Design for Six Sigma. Methodologies that have been widely used for DFSS include Define, Measure, Analyze, Design, Verify (DMADV) and Identify, Design, Optimize, Verify (IDOV). The DMAIC process is used after the product design process has been accomplished and when the product is running in live production. The DFSS process is used during the design of a product or machine so that quality is designed into the product or machine.

DMAIC: Define, Measure, Analyze, Improve, Control—five interconnected phases. Incremental process improvement using Six Sigma methodology. Pronounced Duh-May-Ick, it refers to a data-driven quality strategy for improving processes and is an integral part of the company's Six Sigma quality initiative.

FIFO: First In, First Out.

Gemba: The actual place where customer value is added to the product or service. It is often referred to as the shop floor in a manufacturing environment.

Hard Savings: Six Sigma project benefits that allow you to do the same amount of business with fewer employees (cost savings) or handle more business without adding people (cost avoidance). These are referred to as hard savings. They are the opposite of soft savings.

Jishu Hozen: Autonomous maintenance, a method for employees to take care of small maintenance tasks in their work areas, consequently freeing up time of skilled maintenance employees for more value-added maintenance tasks. The operators are responsible for upkeep of their equipment.

Kaizen: Japanese term that means continuous improvement, taken from the words *Kai*, meaning continuous, and *zen*, which means improvement. Some translate *Kai* to mean change and *zen* to mean good, or for the better. The same Japanese words *Kai* and *zen* are pronounced as *Gai* and *San* in Chinese

Gai: The action to correct.

San: This word is more related to the Taoism or Buddhism philosophy, which give the definition as the action that benefits the society but not one particular individual. The quality of benefit involved here should be sustained forever; in other words, the *san* is an act that truly benefits others.

Kaizen Event: Any action whose output is intended to be an improvement to an existing process. A Kaizen event is commonly referred to as a tool that

1. Gathers operators, managers, and owners of a process in one place
2. Maps the existing process (using a deployment flowchart, in most cases)
3. Improves on the existing process
4. Solicits buy-in from all parties related to the process

Kaizen events are an extremely efficient way to quickly improve a process with a low Sigma score. Kaizen events are also useful for convincing organizations new to Six Sigma of the methodology's value. The true intent of a Kaizen event is to hold small events attended by the owners and operators of a process to make improvements to that process that are within the scope of the process participants.

Kanban: A Japanese term. The actual term means signal. It is one of the primary tools of a Just-in-Time (JIT) manufacturing system. It signals a cycle of replenishment for production and materials. This can be considered as a demand for product from one step in the manufacturing or delivery process to the next. It maintains an orderly and efficient flow of materials throughout the entire manufacturing process with low inventory and work in process. It is usually a printed card that contains specific information such as part name, description, quantity, etc. In a Kanban manufacturing environment, nothing is manufactured unless there is a "signal" to manufacture. This is in contrast to a push-manufacturing environment, where production is continuous.

Little's Law: The fundamental long-term relationship between work in progress, throughput, and flow time of a production system in steady state: Inventory = Throughput × Flow Time.

Pacemaker Process: Any process along a value stream that sets the pace for the entire stream. (The Pacemaker Process should not be confused with a bottleneck process, which necessarily constrains downstream processes due to a lack of capacity.) The pacemaker process is usually near the customer end of the value stream, often the final assembly cell. It is the point of entry of the customer's order into the Lean value stream. In other words, the Pacemaker Process is the process in the Lean value stream that is directly controlled by the customer's order. The Pacemaker Process sets the "pace" for all processes upstream of it in the Lean value stream. All output from the Pacemaker Process flows directly to the customer. There are no pull systems (definition below) or supermarkets (definition below) downstream of the Pacemaker Process. The lead time for the output of a Lean value stream process using a supermarket pull system is the difference between the time the customer order is received by the Pacemaker Process and the time finished output is placed in the supermarket before the Pacemaker Process.

Poka-Yoke: A Japanese term that means mistake-proofing. A poka-yoke device is one that prevents incorrect parts or human errors from being assembled or easily identifies a flaw or error made.

Setup Time: The time interval between the last part from an operation until the first good part is run, subsequent to a model or option changeover. (Note: The terms *setup* and *changeover* are synonymous.)

Single-Focus Strategy: Too many organizations have a plethora of initiatives and metrics. When this happens, improvement resources are spread thin, and because of this, results are achieved at a slower pace. By selecting a single focus or direction, everyone is clear on the direction, resources are focused where they need to be, and everyone understands how to contribute. Because everyone is involved and understands how to contribute, they also automatically improve other areas where improvement is needed. Having a single focus does not mean you abandon everything else but rather that everyone is much clearer on the main direction.

SIPOC: Suppliers, Inputs, Process, Outputs, Customers.

SMED: Single-Minute Exchange of Dies, an improvement program developed by Toyota to reduce the time it takes to set up a process step.

Soft Savings: Six Sigma project benefits such as reduced time to market, cost avoidance, lost profit avoidance, improved employee morale, enhanced image for the organization, and other intangibles may result in additional savings to an organization but are harder to quantify. These are referred to as soft savings. They are different from hard savings.

Spider: In a Lean context, a spider is a person assigned to support a production operation so that others may focus exclusively on value-added. They are also sometimes called a water spider.

Standard Deviation: A statistic used to measure the variation in a distribution. Sample standard deviation is equal to the square root of (the sum of the squared deviations of the mean divided by the sample size minus 1). Where the whole population is known, the –1 "fudge factor" should be omitted. This fudge factor is degrees of freedom. Therefore, as the size of the population increases, the impact of the –1 fudge factor decreases. For a very small sample size, this –1 fudge factor can be significant.

Standard deviation can be thought of as the average distance each data point is from the average of the entire data set. That is, add the squared distance of each point from the average value, and divide by the number of points in the population, or 1 minus the number of data points for a sample, then take the square root of the answer. The distances are squared to eliminate any negative values. It is the most common measure of the variability of a set of data. If the standard deviation is based on a sampling, it is referred to as s. When describing the population, the lowercase Greek letter sigma is used. Although it is closely related to, and used in calculations for the Sigma level of a process, you need to be careful to distinguish the two meanings

STP: Support, Teach, and Promote. A technique used during the manager's daily Gemba walk to support continuous improvement efforts, to teach when people do not quite understand a Lean Sigma tool or new solution going into place, or to promote the single-focus strategy and the culture the organization is seeking.

Supermarket: An inventory facility (container) that is used if continuous flow does not extend upstream in a process; in other words, if batching is necessary, then supermarkets are used to regulate process flows. Withdrawal cards and Kanbans are used for the regulation.

Takt Time: The available production time divided by customer demand. For example, if a widget factory operates 480 minutes per day and customers demand 240 widgets per day, Takt time is two minutes. The purpose of Takt time is to precisely match production with demand. It provides the heartbeat of a Lean production system. Takt time was first used as a production management tool in the German aircraft industry in the 1930s. (*Takt* is German for a precise interval of time such as a musical meter.) It was the interval at which aircraft were moved ahead to the next production station. The concept was widely utilized within Toyota in the 1950s and was in widespread use throughout the Toyota supply base by the late 1960s. Toyota typically reviews the Takt time for a process every month, with a tweaking review every 10 days.

TPM: Total Productive Maintenance is a theory useful for maintaining plants and equipment with total involvement from all employees. Its objectives are to dramatically increase production and employee morale by (1) decreasing waste, (2) reducing costs, (3) decreasing batch sizes, (4) increasing production velocity, and (5) increasing quality (conforming goods). It has three components—autonomous maintenance, predictive maintenance, and preventive maintenance—to meet these objectives.

TWI-JI: Training Within Industry-Job Instruction is designed to develop basic stability of your processes (standard work). This program teaches the method to instruct an operator how to perform a job correctly, safely, and conscientiously. There are three other TWI programs: TWI-JR (Job Relations), TWI-JM (Job Methods), and TWI-JS (Job Safety). All are excellent programs when working with self-directed teams. You can find out more at http://twi-institute.com/.

Value Stream: All of the actions, both value-creating and non-value-creating, required to bring a product from concept to launch and from order to delivery. These include actions to process information from the customer and actions to transform the product on its way to the customer.

Value Stream Map: Value stream mapping is a Lean manufacturing or Lean enterprise technique used to document, analyze, and improve the flow of information or materials required to produce a product or service for a customer. Value stream mapping is a paper and pencil tool (can also be software) that helps you to see and understand the flow of material and information as a product or service makes its way through the value stream. Value stream mapping is typically used in Lean; it differs from the process mapping (flow charting) of Six Sigma in four ways:

1. It gathers and displays a far broader range of information than a typical process map.
2. It tends to be at a higher level (5–10 boxes) than many process maps.
3. It tends to be used at a broader level, i.e., from receipt of raw material to delivery of finished goods.
4. It tends to be used to identify where to focus future projects, subprojects, and/or Kaizen events.

A value stream map (or end-to-end system map) takes into account not only the activity of the product but the management and information systems that support the basic process. This is especially helpful when working to reduce cycle time, because you gain insight into the decision-making flow in addition to the process flow. It is actually a Lean tool. The basic idea is to first map your process, then above it, map the information flow that enables the process to occur.

Index

Note: Page numbers ending in "f" refer to figures. Page numbers ending in "t" refer to tables.